T0302214

Advanced Ceramic Coatings and Interfaces II

Advanced Ceramic Coatings and Interfaces II

A Collection of Papers Presented at the 31st International Conference on Advanced Ceramics and Composites
January 21–26, 2007
Daytona Beach, Florida

Editors
Uwe Schulz
Hua-Tay Lin

Volume Editors
Jonathan Salem
Dongming Zhu

WILEY-INTERSCIENCE
A John Wiley & Sons, Inc., Publication

Published by John Wiley & Sons, Inc., Hoboken, New Jersey.
Published simultaneously in Canada.

For general information on our other products and services or for technical support, please contact our Customer Care Department within the United States at (800) 762-2974, outside the United States at (317) 572-3993 or fax (317) 572-4002.

Wiley also publishes its books in a variety of electronic formats. Some content that appears in print may not be available in electronic format. For information about Wiley products, visit our web site at www.wiley.com.

Wiley Bicentennial Logo: Richard J. Pacifico

Library of Congress Cataloging-in-Publication Data is available.

ISBN 978-0-470-19634-2

Printed in the United States of America.

10 9 8 7 6 5 4 3 2 1

Contents

ENGINEERING OF THERMAL PROPERTIES OF THERMAL BARRIER COATINGS

COATINGS TO RESIST WEAR, EROSION, AND TRIBOLOGICAL LOADINGS

COATINGS FOR SPACE APPLICATIONS

MULTIFUNCTIONAL COATINGS, NANOSTRUCTURED COATINGS, AND INTERFACES PHENOMENA

Preface

The Symposium on Advanced Ceramic Coatings for Structural, Environmental, and Functional Applications was held at the 31st International Conference on Advanced Ceramics and Composites, January 21-26, 2007, Daytona Beach, Florida. A total of 85 papers including 14 invited papers were presented at the symposium focusing the latest advances in research and development and applications of coating technologies.

This proceedings contains 24 contributed papers from the symposium, covering topics of thermal barrier coatings, coatings to resist wear, erosion, and tribological loadings, coatings for space applications, and multifunctional, nanostructured coatings, and interfaces phenomena. These papers provide the state-of-art of ceramic coating technologies for various critical industrial technology applications.

We are greatly indebt to the members of the symposium organizing committee, including Drs. Yutaka Kagawa, Anette Karlsson, Seiji Kuroda, Karren More, Jennifer Sample, Dileep Singh, Yong-Ho Sohn, Irene Spitsberg, Robert Vaßen, and Dongming Zhu, for their tremendous time and effort in developing and organizing this vibrant and cutting-edge symposium. We also would like to express our sincere thanks to all of the symposium participants, session chairs, manuscript authors, and reviewers, for their contributions to such a successful and excellent quality meeting. Finally, we are also very grateful to The American Ceramic Society staff for their help during the organization and publication of this symposium and proceedings.

UWE SCHULZ
German Aerospace Center

HUA-TAY LIN
Oak Ridge National Laboratory

Introduction

2007 represented another year of growth for the International Conference on Advanced Ceramics and Composites, held in Daytona Beach, Florida on January 21-26, 2007 and organized by the Engineering Ceramics Division (ECD) in conjunction with the Electronics Division (ED) of The American Ceramic Society (ACerS). This continued growth clearly demonstrates the meetings leadership role as a forum for dissemination and collaboration regarding ceramic materials. 2007 was also the first year that the meeting venue changed from Cocoa Beach, where it was originally held in 1977, to Daytona Beach so that more attendees and exhibitors could be accommodated. Although the thought of changing the venue created considerable angst for many regular attendees, the change was a great success with 1252 attendees from 42 countries. The leadership role in the venue change was played by Edgar Lara-Curzio and the ECD's Executive Committee, and the membership is indebted for their effort in establishing an excellent venue.

The 31st International Conference on Advanced Ceramics and Composites meeting hosted 740 presentations on topics ranging from ceramic nanomaterials to structural reliability of ceramic components, demonstrating the linkage between materials science developments at the atomic level and macro level structural applications. The conference was organized into the following symposia and focused sessions:

- Processing, Properties and Performance of Engineering Ceramics and Composites
- Advanced Ceramic Coatings for Structural, Environmental and Functional Applications
- Solid Oxide Fuel Cells (SOFC): Materials, Science and Technology
- Ceramic Armor
- Bioceramics and Biocomposites
- Thermoelectric Materials for Power Conversion Applications
- Nanostructured Materials and Nanotechnology: Development and Applications
- Advanced Processing and Manufacturing Technologies for Structural and Multifunctional Materials and Systems (APMT)

- Porous Ceramics: Novel Developments and Applications
- Advanced Dielectric, Piezoelectric and Ferroelectric Materials
- Transparent Electronic Ceramics
- Electroceramic Materials for Sensors
- Geopolymers

The papers that were submitted and accepted from the meeting after a peer review process were organized into 8 issues of the 2007 Ceramic Engineering & Science Proceedings (CESP); Volume 28, Issues 2-9, 2007 as outlined below:

- Mechanical Properties and Performance of Engineering Ceramics and Composites III, CESP Volume 28, Issue 2
- Advanced Ceramic Coatings and Interfaces II, CESP, Volume 28, Issue 3
- Advances in Solid Oxide Fuel Cells III, CESP, Volume 28, Issue 4
- Advances in Ceramic Armor III, CESP, Volume 28, Issue 5
- Nanostructured Materials and Nanotechnology, CESP, Volume 28, Issue 6
- Advanced Processing and Manufacturing Technologies for Structural and Multifunctional Materials, CESP, Volume 28, Issue 7
- Advances in Electronic Ceramics, CESP, Volume 28, Issue 8
- Developments in Porous, Biological and Geopolymer Ceramics, CESP, Volume 28, Issue 9

The organization of the Daytona Beach meeting and the publication of these proceedings were possible thanks to the professional staff of The American Ceramic Society and the tireless dedication of many Engineering Ceramics Division and Electronics Division members. We would especially like to express our sincere thanks to the symposia organizers, session chairs, presenters and conference attendees, for their efforts and enthusiastic participation in the vibrant and cutting-edge conference.

ACerS and the ECD invite you to attend the 32nd International Conference on Advanced Ceramics and Composites (http://www.ceramics.org/meetings/daytona2008) January 27 - February 1, 2008 in Daytona Beach, Florida.

JONATHAN SALEM AND DONGMING ZHU, Volume Editors
NASA Glenn Research Center
Cleveland, Ohio

Thermal and Mechanical Properties of Thermal Barrier Coatings

THERMAL AND MECHANICAL PROPERTIES OF ZIRCONIA COATINGS PRODUCED BY ELECTROPHORETIC DEPOSITION

Bernd Baufeld & Omer van der Biest
Metaalkunde en Toegepaste Materiaalkunde
Katholieke Universiteit Leuven
Kasteelpark Arenberg 44
3001 Leuven, Belgium

Hans-Joachim Rätzer-Scheibe
Institute of Materials Research
German Aerospace Center (DLR)
Linder Höhe
51147 Cologne, Germany

ABSTRACT

The topic of this paper is electrophoretic deposition (EPD) as a cheap and fast coating procedure to obtain thermal barrier coatings. In EPD a coating is obtained by deposition of powder from a suspension under the influence of an electric field and a subsequent consolidation by sintering. Crack free, up to 0.15 mm thick coatings with homogenous morphology and high porosity were obtained.

The elastic modulus of an EPD coating was determined to be 22 GPa at room temperature decreasing to 18 GPa at 1000°C. The thermal conductivity has, depending on porosity, values between 0.5 and 0.6 W/(m·K) at room temperature, which decrease slightly with temperature. After annealing in air at 1100°C for 100 h the thermal conductivity has been increased by about 50 %.

The low elastic modulus and the exceptionally low thermal conductivity, both related to the high porosity, make the EPD coatings a promising candidate for thermal barrier coatings.

INTRODUCTION

The thermal barrier coatings (TBC) for gas turbine engine applications usually consist of partially yttria stabilized zirconia due to its low thermal conductivity and its relatively high thermal expansion coefficient. The most common techniques for depositing TBCs are either air plasma spraying (APS) or electron beam physical vapor deposition (EB-PVD), the first mostly applied for energy transformation and the latter for aeronautics. In the as-coated condition the APS coatings have at room temperature thermal conductivity values in the range of 0.6–1.4 W/(m·K)[1-5], while the EB-PVD coatings usually have higher values of 1.5–2.0 W/(m·K)[1, 4-7]. However, these reliable and established coating techniques are cost and time intensive. As a cheaper and faster coating procedure, the use of electrophoretic deposition (EPD) was suggested[8].

EPD is a colloidal deposition process, where in a first step the powder in a suspension is deposited under the influence of an electric field on an electrode and then, in a second step, the coating is consolidated by a heat treatment. A wide range of metallic as well as ceramic powders are already studied for EPD, either in aqueous or organic liquids[9-11].

3

Until now, not much work has been performed in investigating the material properties of ceramic coatings fabricated by EPD. A few authors report about hardness and elastic modulus of EPD coatings and composites, determined with indenter techniques[12-15]. Preliminary results concerning elastic modulus and damping derived from impulse excitation technique (IET) are published by some of the present authors[16]. Reports about the thermal conductivity of ceramic EPD coatings were not found in the open literature.

Yet, for the application of ceramic coatings prepared by EPD these parameters are essential. In this paper results about thermal conductivity measured by the laser flash method and elastic modulus derived from IET will be presented in dependence on the temperature.

EXPERIMENTAL PROCEDURE
Specimen preparation

Partially yttria stabilized zirconia powder (5 mol% Y_2O_3 stabilized grade Melox 5Y XZO 99.8%) with the addition of 0.75 wt% cobalt oxide nanopowder as a sintering aid (Aldrich Cobalt (II, III) oxide 99.8%) was used. Two different suspensions were investigated, one methyl-ethyl-ketone (MEK) based and one ethanol based. The powder was mixed and ball milled with zirconia balls in MEK or in ethanol with a multidirectional mixer (Turbula type) for one day. For each EPD session fresh suspensions were prepared by adding n-butylamine (BA) and a respective suspension stabilization additive (see Tab. I). The suspensions were first mechanically, then ultrasonically, and finally again mechanically stirred, for 15 minutes each sequence.

The EPD cells consisted of non-conductive containers with the substrate as one electrode. The specimens for the laser flash experiment were placed in a cell with vertical plane electrodes, while the specimen for the IET was installed horizontally in the center of a cylindrical counter-electrode. More details of the latter cell can be found elsewhere[16]. During EPD, the suspension was subjected in both cases to further mechanical stirring. The experimental conditions and the results are presented in Tab. I.

For the laser flash method a special type of sample was used, which has been called quasi-free-standing coating[4], and allows to measure the thermal conductivity of fragile and thin coatings. This sample type consists of a sapphire support (diameter 12.7 mm, thickness 1 mm) on which the ceramic EPD coating is applied (Fig. 1). For the laser flash measurement of semi-transparent zirconia it is necessary to add a thin platinum layer on both sides of the ceramic coating, which were sputter coated before and after the EPD process with a thickness of about 5 μm. At the front surface this Pt layer prevents the laser beam penetration into the interior of the sample and ensures an effective and uniform absorption of the laser pulse. The thin platinum

Tab. I Conditions and results of the EPD (for the porosity a density of 6.05 g/cm^3 of fully dense zirconia[17] was assumed)

Specimen name	EPD1	EPD2	EPD3
Test type	IET	Laser flash	Laser flash
Electrical field strength [V/mm]	17	8.6	2.8
Suspension based on	MEK (20 vol% BA)		Ethanol (3 vol% BA)
Suspension additives	1 wt% nitro-cellulose		1 wt% Dolapix
Powder load	63 g/l		96 g/l
EPD time [s]	50	120	120
EPD coating thickness [mm]	0.15	0.11	0.10
porosity	0.42	0.47	0.38

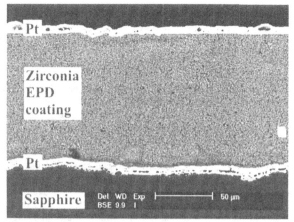

Fig. 1 Cross section of a quasi-free standing EPD coating (specimen EPD2).

layer on the rear EPD surface between sapphire and coating serves as emission layer for measuring the temperature signal by the infrared detector. In addition, this opaque layer prevents the infrared detector from viewing into the sample interior and thus not giving an accurate temperature rise curve for the rear surface. Furthermore, for the EPD process the metal layer on the electrical nonconductive sapphire forms the necessary electrode.

For the coating of an IET specimen, a cylindrical rod of polycrystalline Ni based superalloy IN625 with a length of 65 mm and a diameter of 6 mm was used.

After EPD, the containers were drained and the specimens dried in air. The green specimens were sintered in a conventional resistance furnace in hydrogen atmosphere at 1200°C for 6 h. The density of the EPD coating was determined by geometrical means.

For the IET experiments, the EPD coated specimen was cut to a length of 47 mm and a second, uncoated IN625 specimen (diameter 6 mm) to a length of 54 mm.

Impulse Excitation Technique

IET is a standard test method for the assessment of the dynamic elastic modulus of isotropic homogeneous materials[18]. Basis of IET is the stimulation of the resonance frequency by gently hitting it, and analyzing the resultant vibration signal. The elastic modulus is proportional to the square of the resonance frequency and depends furthermore on mass, geometry and Poisson's ration of the specimen. In the case of coated cylindrical specimens it has been shown by Schrooten et al.[19], that the elastic modulus of the coating E_c can be expressed as a function of the total elastic moduli in the flexural mode of the coated system E_{tot} and of the substrate E_s, of the radius R of the substrate and of the thickness of the coating t:

$$E_C = \frac{E_{tot}(R+t)^4 - E_s R^4}{(R+t)^4 - R^4} \quad (1)$$

Therefore, in order to obtain the elastic modulus of the coating two specimens, one with and one without coating, have to be tested.

The IET tests were performed in a graphite furnace HTVP-1750C in Ar atmosphere up to 1000°C, with a heating and cooling rate of 2°C/min, and analyzed by the RFDA software (both IMCE, Diepenbeek, Belgium) resulting in elastic modulus and damping in dependence of the temperature. More details about this set up can be found elsewhere[20].

Laser flash technique

The thermal diffusivity of the EPD coating was measured using the laser flash method (LFA427, NETZSCH). In this technique, the front side of a plan-parallel disk is heated by a short laser pulse (neodymium doped gallium gadolinium garnet laser with a wavelength of 1.064 µm). The heat diffuses through the sample and the resulting temperature rise of the rear surface is recorded by an infrared detector (2 to 5 µm wavelength range). The thermal diffusivity is determined from the measured temperature rise and analyzed by the NETZSCH LFA Proteus® software. Thermal conductivity of the coating k was then calculated using the relationship k $=c_p\rho\alpha$, where c_p is the temperature dependent specific heat[21], ρ the density, and α the thermal diffusivity. The analysis is based on a one-dimensional heat flow and takes the heat loss from all surfaces of the disk-shaped sample into account. In contrast to former analyses[4], the effect of the radiative heat transfer in the interior of the sample, which is especially noticeable at higher temperatures[22], was considered by this software.

The tests were performed in vacuum or in Ar gas at atmospheric pressure from room temperature up to 1150°C. In order to study sintering effects on the thermal conductivity, which are reported for conventional TBC systems[3], specimen EPD3 was not only investigated in the as-received condition, but also after 100 h annealing in air at 1100°C.

RESULTS

Electrophoretic deposition

Under the chosen conditions up to 0.15 mm thick coatings with smooth surface and without visible cracks were obtained (Tab. I). The coatings have a high porosity with evenly distributed

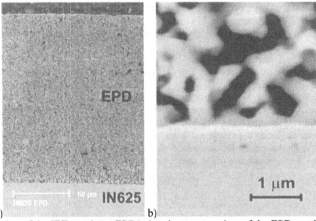

a)   b)

Fig. 2 Cross section of the IET specimen EPD1 showing an overview of the EPD coating (a) and in detail the interface between EPD coating and substrate (b).

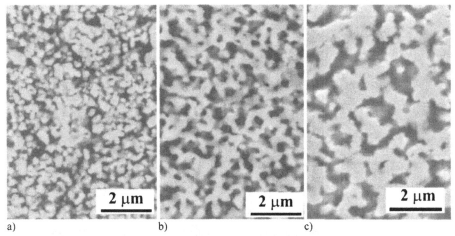

a)   b)   c)

Fig. 3 Microstructure of EPD2 (a) and EPD3 (b) in the as-received state, and EPD3 after 100 h heat treatment at 1100°C in air (c).

pores and some agglomerates (Fig. 1,Fig. 2, Fig. 3a and b). The adherence of the coating is sufficient to allow, for example, mechanical cutting of a specimen. For EPD1 a compressive residual stress of (-91 ± 17) MPa was measured, using XRD. The 100h heat treatment at 1100°C resulted in a significant microstructural change of the coating with increased grain and pore sizes (Fig. 3c).

Impulse excitation technique

The elastic modulus of the coated specimen and of the substrate alone was determined for room temperature to be 186 GPa and 209 GPa, respectively. According to equation (1), the elastic modulus of the EPD coating is calculated to be 22 GPa. With increasing temperature the elastic modulus of IN625 and of the coated system decreases significantly (Fig. 4a). The elastic modulus of the coating, however, decreases only slightly to 18 GPa at 1000°C. While the elastic modulus of the EPD coating is much smaller than the one of dense zirconia, it is in the same order as reported for air plasma sprayed[23] or EB-PVD zirconia coatings[24].

For the coated as well as for the uncoated specimen the damping increases with temperature (Fig. 4b). At temperatures below 850°C, however, the damping of the coated specimen is significantly higher than for the uncoated specimen, supposedly due to energy dissipating processes within the porous ceramic coating. Such an increased damping is a beneficial property, since this may reduce vibration or noise in a turbine. For example, air plasma sprayed zirconia coatings designed for this task were studied by Yu et. al.[25].

Laser flash method

The thermal conductivity of the EPD coatings prepared from different suspensions are at room temperature roughly 0.5 and 0.6 W/(m·K) for the MEK based and the ethanol based EPD coating, respectively, decreasing with temperature (Fig. 5a). Due to the lower porosity the MEK suspension resulted in a coating with slightly lower thermal conductivity than the one of the

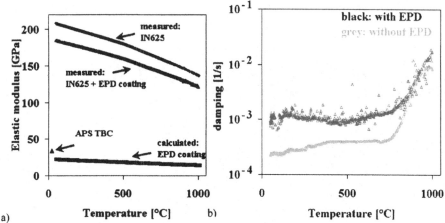

Fig. 4 a: Temperature dependence of the elastic modulus of the coated system, the uncoated substrate, and of the coating. For comparison, the elastic modulus of an APS zirconia coating is given[23]. b: Temperature dependence of the damping of a coated and an uncoated specimen.

ethanol based suspension. The testing atmosphere has only a minor importance with slightly higher thermal conductivity at lower temperatures in Ar atmosphere compared to vacuum (Fig. 5a).

In comparison with typical APS or EB-PVD coatings (Fig. 5b), the as-received EPD coatings have a remarkably low thermal conductivity. However, annealing at 1100°C for 100 h in air substantially increased the thermal conductivity by about 50 %. The thermal conductivity is still lower than for commonly used TBC systems, for which also the thermal conductivity

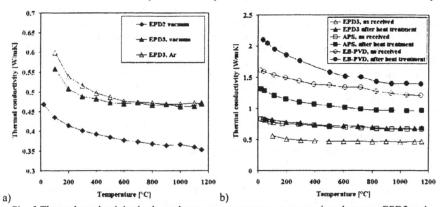

Fig. 5 Thermal conductivity in dependence on temperature. a: comparison between EPD2 and EPD3 in vacuum and in Ar atmosphere. b: comparison of EPD, APS and EB-PVD zirconia coatings in vacuum for as-received specimens and after 100 h annealing in air at 1100°C.

increases significantly (Fig. 5b). Details about the EB-PVD and APS measurements can be found elsewhere[5]. It is worthwhile to note, that in contrast to previously reported results[4], the thermal conductivity of the EB-PVD coating (0.280 mm thick) does not increase at higher temperatures. This is due to the fact, that in the present work the data were corrected by taking the contribution of radiation to the heat conduction[22] into account.

CONCLUSIONS

Crack free, but porous ceramic EPD coatings of more than 0.1 mm thickness have been obtained by MEK, as well as by ethanol based suspensions. The mechanical properties of the coatings were studied with IET resulting in a low elastic modulus of 22 GPa decreasing only slightly with increasing temperature. The EPD coating increased the damping properties of the coated system. The thermal conductivity of the EPD coating proved to be remarkably low. Annealing at 1100°C, however, increased the thermal conductivity by about 50 %.The thermal and mechanical properties of the EPD coatings are very promising for the application as TBC. However, adherence, especially for thicker coatings, is an issue to be studied in future.

ACKNOWLEDGMENTS

B. Baufeld acknowledges an individual Marie-Curie fellowship of the European Commission Nr. MEIF-CT-2005-010277.

REFERENCES

[1]R. B. Dinwiddie, S. C. Beecher, W. D. Porter and B. A. Nagaraj, "The effect of thermal aging on the thermal conductivity of plasma sprayed and EB-PVD thermal barrier coatings", in *Turbo Expo '96* (ASME, Birmingham UK, 1996).

[2]A. J. Slifka, B. J. Filla, J. M. Phelps, G. Bancke and C. C. Berndt, "Thermal conductivity of a zirconia thermal barrier coating", *J. Therm. Spr. Tech.*, **7**, 43-46 (1998).

[3]A. A. Kulkarni, A. Vaidya, A. Goland, S. Sampath and H. Herman, "Processing effects on porosity-property correlations in plasma sprayed yttria-stabilized zirconia coatings", *Mat. Sci. Eng. A*, **359**, 100-11 (2003).

[4]H. J. Rätzer-Scheibe, U. Schulz and T. Krell, "The effect of coating thickness on the thermal conductivity of EB-PVD PYSZ thermal barrier coatings", *Surf. Coat. Tech.*, **200**, 5636-44 (2006).

[5]H. J. Rätzer-Scheibe and U. Schulz, "The effect of heat treatment and gaseous atmosphere on the thermal conductivity of APS and EB-PVD thermal barrier coatings", *Surf. Coat. Tech.*, submitted, (2007).

[6]K. An, K. S. Ravichandran, R. E. Dutton and S. L. Semiatin, "Microstructure, texture, and thermal conductivity of single-layer and multilayer thermal barrier coatings of Y2O3-stabilized ZrO2 and Al2O3 made by physical vapor deposition", *J. Am. Ceram. Soc.*, **82**, 399-406 (1999).

[7]J. R. Nicholls, K. J. Lawson, A. Johnstone and D. S. Rickerby, "Low Thermal Conductivity EB-PVD Thermal Barrier Coatings", *Mat. Sci. For.*, **369-372**, 595-606 (2001).

[8]O. van der Biest, E. Joost, J. Vleugels and B. Baufeld, "Electrophoretic deposition of zirconia layers for thermal barrier coatings", *J. Mat. Sci.*, **41**, 8086-8092 (2006).

[9]O. Van der Biest and L. Vanderperre, "Electrophoretic deposition of materials", *Annu. Rev. Mater. Sci.*, **29**, 327-52 (1999).

[10]A. R. Boccaccini and I. Zhitomirsky, "Application of electrophoretic and electrolytic deposition techniques in ceramics processing", *Cur. Op. Sol. St. Mat. Sci.*, **6**, 251-60 (2002).

[11]L. Besra and M. Liu, "A review on fundamentals and applications of electrophoretic deposition (EPD)", *Prog. Mat. Sci.*, **52**, 1-61 (2007).

[12]X. Wang, P. Xiao, M. Schmidt and L. Li, "Laser processing of yttria stabilised zirconia/alumina coatings on Fecralloy substrates", *Surf. Coat. Tech.*, **187**, 370-76 (2004).

[13]S. Put, J. Vleugels, G. Anne and O. Van der Biest, "Functionally graded ceramic and ceramic-metal composites shaped by electrophoretic deposition", *Coll. Surf. A*, **222**, 223-32 (2003).

[14]X.-J. Lu, X. Wang and P. Xiao, "Nanoindentation and residual stress measurements of yttria-stablized zirconia composite coatings produced by electrophoretic deposition", *Thin Solid Films*, **494**, 223-27 (2006).

[15]P. Hvizdos, J.-M. Calderon Moreno, J. Ocenasek, L. Ceseracciu and G. Anne, "Mechanical Properties of Alumina/Zirconia Functionally Graded Material Prepared by Electrophoretic Deposition", *Key. Eng. Mat.*, **290**, 332-35 (2005).

[16]B. Baufeld and O. van der Biest, "Development of thin ceramic coatings for the protection against temperature and stress induced rumpling of the metal surface of turbine blades", *Key. Eng. Mat.*, **333**, 273-76 (2007).

[17]R. P. Ingel and D. I. Lewis, "Lattice parameters and density for Y_2O_3-stabilized ZrO_2", *J. Am. Ceram. Soc.*, **69**, 325-32 (1986).

[18]ASTM, "Standard Test Method for Dynamic Young's Modulus, Shear Modulus, and Poisson's Ration by Impulse Excitation of Vibration", **E 1876-99**, 1075-84 (1999).

[19]J. Schrooten, G. Roebben and J. A. Helsen, "Young's modulus of bioactive glass coated oral implants: porosity corrected bulk modulus versus resonance frequency analysis", *Scr. Mat.*, **41**, 1047-53 (1999).

[20]G. Roebben, B. Bollen, A. Brebels, J. Van Humbeeck and O. Van der Biest, "Impulse excitation apparatus to measure resonant frequencies, elastic moduli and internal friction at room and high temperature", *Rev. Sci. Instrum.*, **68**, 4511-15 (1997).

[21]T. Krell, "Thermische und thermophysikalische Eigenschaften von elektronenstrahl-gedampften chemisch gradierten Al_2O_3/PYSZ-Wärmedämmschichten", Dissertation RWTH Aachen, (2000).

[22]F. Schmitz, D. Hehn and H. R. Maier, "Evaluation of laser-flash measurements by means of numerical solutions of the heat conduction equation", *High Temp., High Pres.*, **31**, 203-11 (1999).

[23]J. S. Wallace and J. Ilavsky, "Elastic Modulus Measurements in Plasma Sprayed Deposits", *J. Therm. Spr. Tech.*, **7**, 521-26 (1998).

[24]U. Schulz, K. Fritscher, C. Leyens and M. Peters, "Influence of processing on microstructure and performance of EB-PVD thermal barrier coatings", *J. Eng. Gas Turb. Pow.*, **124**, 1-8 (2000).

[25]L. Yu, Y. Ma, C. Zhou and H. Xu, "Damping capacity and dynamic mechanical characteristics of the plasma-sprayed coatings", *Mat. Sci. Eng. A*, **408**, 42-46 (2005).

ELASTIC AND INELASTIC DEFORMATION PROPERTIES OF FREE STANDING CERAMIC EB-PVD COATINGS

Marion Bartsch and Uwe Fuchs
Institute of Materials Research, German Aerospace Center (DLR)
Linder Hoehe
D-51103 Cologne, Germany

Jianmin Xu
Rolls-Royce Deutschland Ltd & Co KG
D-15827 Blankenfelde-Mahlow, Germany

ABSTRACT

Thermal barrier coatings (TBC) processed by electron beam physical vapor deposition (EB-PVD) show complex deformation behavior due to their columnar microstructure and multi-scale porosity. Limited information is available regarding the inelastic deformation behavior. Furthermore, the existing data for the elastic properties of EB-PVD coatings varies in a wide range, depending on the measurement method. In most cases, the fragile EB-PVD coatings are tested on the substrate, onto which they were deposited. Since the ceramic coatings are deposited onto metallic substrates, which have about 1000°C, compressive residual stresses develop in the coating after cooling down. Since the deformation behavior of EB-PVD coatings is stress dependent, the residual stresses cannot be calculated without knowledge of the stress-strain response, and without knowledge of the residual stress the elastic properties cannot be measured correctly. To overcome this dilemma, tubular free standing EB-PVD coating samples with a thickness of 250µm have been processed and tested in compression at room temperature. For introducing the mechanical load safely into the gauge length a soft clamping has been developed, and the strain was measured using a contact-free laser extensometer. Several subsequent loading and unloading cycles were performed on specimens with different thermal pre heat treatments. Non-linear stress-strain behavior was observed, showing increasing stiffness with increasing compressive stress. Cycle by cycle the stiffness increased, indicating inelastic deformation, which has also been observed in creep experiments in the same test configuration. Long term thermally aged specimens showed higher stiffness than as processed specimens.

INTRODUCTION

The knowledge of elastic and inelastic deformation properties is important to understand and model the damage and failure behavior of EB-PVD thermal barrier coatings. The existing data on EB-PVD TBCs are limited and varies in a wide range, depending on the testing method. Several authors observed evidence for stress dependence of the elastic properties of EB-PVD coatings[1,2,3,4]. This is important, since EB-PVD coatings develop high compressive residual stresses after deposition on metallic substrates after cooling down due to the thermal mismatch between substrate and coating. Thus, it is necessary to know the residual stress state when determining the elastic properties of the EB-PVD coating. Since a free standing EB-PVD TBC is nearly stress free, it would be ideal to perform mechanical tests on it, provided that it would not be so extremely fragile. The deposition of significantly thicker TBCs does not help solving the problem since the mechanical properties of EB-PVD coatings depend on the thickness due to a porosity gradient over the thickness, which corresponds to the difference between near-to-interface and distant-to-interface columnar microstructures. For example, finite element calculations, considering the measured inter-columnar porosity over the thickness of an EB-PVD coating, predict a decrease in elastic modulus of about 30% from the near-to-interface to a 60 μm distant-to-interface microstructure [5]. The porosity gradient has also to be considered when testing flat free standing specimens in bending. The challenge is to achieve free standing coatings like for industrial applications with a thickness of about 200 to 300 μm with sufficient stability and to test them in-plane. Some advantages concerning stability provide tubular specimens, which have been successfully processed in the Institute of Materials Research of the German Aerospace Center in Cologne. A technique has been developed by the authors to test these tubular specimens in compression, using soft fixtures to clamp the specimens and a laser extensometer for contact-free strain measurement. Tests have been performed on specimens, which have been mechanically stabilized by thermal pre-treatment and on specimens, which have been additionally aged at high temperature, in order to study the effect on changes of the mechanical properties due to the thermal ageing. The data obtained on the free standing tubular specimens are compared to data reported in the literature on the same coating system but using different testing methods.

EXPERIMENTAL
Specimens

EB-PVD coatings were deposited onto metallic cylinder-shaped substrates using standard coating conditions with rotating the cylinders around their length axis. The coating was subsequently removed from the substrate so that as free standing tubes with a thickness of 250μm were produced, which is typical in standard coatings for industrial application. Tubes with 2 different diameters - 12, and 19 mm - were processed. Specimens of about 20mm length were cut from the tubes by laser cutting. The cut edges were plan-parallel but small parts of the coatings always broke from the edges, see fig. 1.

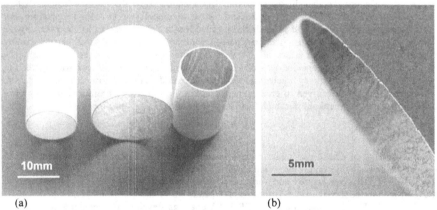

(a)   (b)

Fig.1. Free standing tubular specimens after laser cutting (a) and detail of a laser cut edge (b).

The ceramic coatings were made from zirconia, partially stabilized with 7-8 wt % yttria. The microstructure of the coatings was identical to standard coatings as used for gas turbine blades and showed the typical columnar structure with higher density at the inner surface toward the (later removed) substrate than at the outer surface (fig. 2).

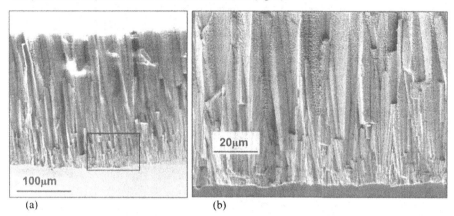

(a)   (b)

Fig. 2. Free standing EB-PVD coating, overview (a) and enlarged detail (b)

Due to their columnar microstructure, the specimens were extremely fragile in the as coated condition. In order to strengthen the connections between the columns and thus to avoid fracture by handling, all specimens were thermally pre treated according to industrial standard. Some specimens have been additionally thermally aged for long term in air in order to investigate the effect of microstructural changes, such as sintering, on the mechanical properties. The long term thermal ageing resulted in a slight reduction of the diameter (12.65 mm to 12.5 mm) and often in damage of the specimens by the formation of cracks at the edges, so that only one intact specimen was available for further testing. An overview of the tested specimens is given in Table I.

Table I: Overview over tested specimens

Specimen Nr.	outer diameter [mm]	thermal ageing in air	comment	performed tests
1	19.5	-		compr. load cycles, creep
2	19,5	-		compr. load cycles, creep
3	19.5	-		1 compressive load cycle
4	12.5	1130°C/ 201h		compr. load cycles
5	12.5	1130°C/ 238h	partially cracked at edge	compressive load cycles

Testing procedure

Before testing in compression, soft fixtures were attached to both ends of the specimens in order to even out slivered parts of the edges, avoid stress concentrations, and stabilize the free standing tubes. The fixtures consist of polycarbonate caps, which were laid out with a layer of flexible glue. The specimens were aligned, the ends fixed to the caps, and glued into the cap. Finally, the specimens were furnished with parallel stripes, which give a good contrast for contact-free strain measurements by means of a laser extensometer in further compression testing. Fig. 3 shows a schematic of the fixture and a photograph of a specimen in testing configuration.

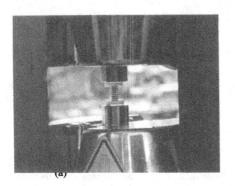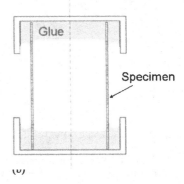

Fig.3. Photograph of a specimen in testing configuration (a) and schematic of the specimen fixture (a)

Compressive loading and unloading tests and some creep tests have been performed in the same test configuration with an electro-mechanic spindle testing machine. The specimens were loaded in several subsequent test cycles until a defined load and unloaded until a load of about 10N. The load was measured by means of a load cell and controlled manually. The displacement rate of the cross head was in the first experiments selected to 0.01mm/min and changed to 0.1mm/min in order to save testing time. The strain was measured by means of a laser extensometer (Fiedler Measurement Systems), which emits a laser beam scanning the specimen. The laser signals, which were reflected from the contrast-rich stripes on the specimens, are utilized to measure the strain; changes between the distances of the stripes due to deformation of the specimen result in changes between the time distances of the maxima of the reflected laser signal. The precision of the measurement depends on the accuracy of the stripe pattern on the specimen. On the first specimens, the stripe pattern was painted manually using a microscope, resulting in stripes with some arbitrary irregularities. Later, it was recognized that the fixture stabilizes the specimens sufficiently to press on the stripe pattern provided by the manufacturer of the laser extensometer. Using the press-on stripes instead of painting the stripes resulted in significantly less scatter of the monitored strain data.

RESULTS

Several test sets with subsequent compressive loading and unloading cycles have been performed on the specimens, see Table I. Specimens without ageing showed non-linear deformation behavior with an increasing 'elastic modulus' with increasing mechanical load. Between loading and unloading a hysteresis occurred. After the first loading/unloading cycle the slope of the load-strain curve became steeper and the hysteresis loop became narrower. After about 5 cycles a steady state was reached. However, when the test set was repeated after some minutes of total unloading, the specimens recovered and showed almost the same load-strain behavior like in the preceding test set. Fig. 4a displays the load-strain data of an exemplary test set and Fig. 4b gives an overview of measured values for the elastic modulus as a function of the applied compressive stress.

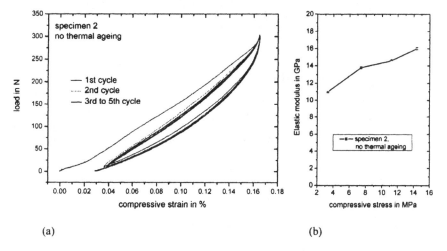

(a)   (b)

Fig. 4. Load – strain behavior of a specimen (a) and overview of measured elastic-moduli as function of compressive stress (b)

Specimens, which have been thermally aged, did not show an expressed non-linear behavior and no hysteresis, but the elastic modulus of the specimens increased in subsequent loading-unloading cycles. Exemplary loading-unloading cycles of specimen 4 and an overview of the measured elastic moduli are displayed in Fig. 5. The specimen 5 developed cracks at the edges during the long term thermal treatment and showed at higher loads untypical non-linear behavior. However, at loads between 50 and 100N the load-strain behavior was almost linear and a value of 66GPa has been measured, which agrees well with the data on the intact thermally aged specimen.

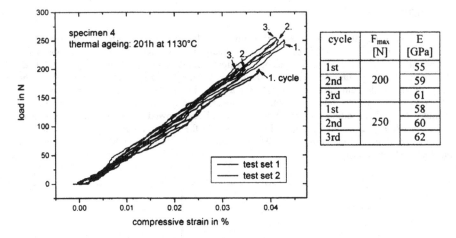

cycle	F_{max} [N]	E [GPa]
1st		55
2nd	200	59
3rd		61
1st		58
2nd	250	60
3rd		62

Fig. 5. Load-strain behavior of a specimen after thermal ageing

Some creep tests have been performed on non aged specimens in the same test configuration like the loading-unloading cycles. The specimens displayed time dependent deformation behavior, specimen Nr. 2 fractured during the creep test after 4900 seconds. The test data are displayed in Fig. 6.

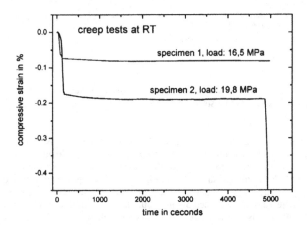

Fig. 6. Compressive strain as function of time under constant load for two specimens without ageing

DISCUSSION

Non linear stress-strain behavior was observed in compressive tests on free standing EB-PVD zirconia coatings. The non linear behavior depends on the stress-level, the loading history, and on the thermal history of the specimens. The stress dependence of EB-PVD coatings is thought to depend on micro-contacts between the columns of the coating, following the micro-contact model discussed by Szücs[2]. In stress free condition micro-contacts between columns ensure the integrity of the coating. With increasing compressive stress, existing micro-contacts deform and subsequently more micro-contacts can form and bridge the gaps between single columns, so that each further deformation increment requires a higher compressive stress increment. This model can be used to explain the non linear load-strain behavior. Assuming that many contacts are not rigid, dissipation by local friction can occur and led to the observed hysteresis between loading and unloading. Thermally aged specimens did not show expressed non-linearity or hystereses, which may be attributed to the formation of many rigid micro-contacts due to sintering. The formation of sinter-contacts between the columns of EB-PVD coatings have been observed by several authors[6,7] and can explain also the higher values obtained for the elastic modulus of thermally aged specimens compared to non aged ones. The micro-contact model may also be used to explain irreversible strain fractions, as indicated by the increase of elastic moduli in subsequent cycles, especially after the first cycle by assuming micro-fracture at the inter-columnar contacts. However, it does not provide a sound explication for the observed recovery effects. Also not clear is the mechanism behind the creep behavior at room temperature. Further experiments and microstructural investigations are planned in order to elucidate the creep mechanisms.

The data for elastic moduli obtained on the free standing tubes vary in the case of the non aged specimens between 10 and 18 GPa in the first load cycle at low stresses of about 4 MPa and between 14 and 26 GPa at about 12 MPa in repeated cycling. The differences between the individual specimens may be attributed to differences in the coating conditions, because the specimens were cut from tubes of about 80 mm length, which are subjected to different coating conditions depending on the location in the vapor cloud during deposition. The obtained values are comparable to the value of 16 GPa, reported by Szücs[2] on 250 μm thick flat free standing specimens in dynamic mechanical bending tests. Data reported on EB-PVD coatings on metallic substrates vary between 15 and 116 GPa, depending on the residual stress in the coating system and the testing method.[1,2,4,8] The thermal ageing led to a significant increase of the elastic modulus of the free standing tubes to values between 55 and 66 GPa. Dynamic mechanical bending tests on 250 μm thick flat free standing specimens gave an increase of the elastic modulus from 16 GPa before to 30.5 GPa after 4h ageing at 1000°C in air[2], and own measurements on 220μm thick EB-PVD coatings on IN 625 substrates resulted in an increase of the elastic modulus from about 35 GPa before to 72 GPa after 1000h ageing at 1000°C in air.[9] In the variety of data the obtained values on the free standing TBC tubes are reasonable.

CONCLUSIONS

Compressive testing of free standing tubular EB-PVD coatings gave reasonable results for the elastic and inelastic deformation behavior. The observed non-linear behavior and loading-unloading hystereses observed on the specimens can be explained by a micro-contact model, assuming the formation of increased inter-columnar contacts during compressive loading.

ACKNOWLEDGEMENTS

This investigation was funded by the German Bundesministerium für Wirtschaft und Technologie on the research program MARCKO with FKZ 03266888C. The authors thank Klaus Kröder and Jörg Brien for manufacturing the free standing tubular coatings, Dan Renusch (DECHEMA e.V) for performing the laser cutting of the specimens, and Liudmila Chernova (German Aerospace Center, DLR) for taking the micrographs.

REFERENCES

[1]C.A. Johnson, J.A. Ruud, R. Bruce, D.Wortmann, "Relationship between residual stress, microstructure and mechanical properties of electron beam – physical vapor deposition thermal barrier coatings", Surf. & Coat. Technology 108-109, 80-85 (1998).

[2]F. Szücs, "Thermomechanische Analyse und Modellierung plasmagespritzter und EB-PVD aufgedampfter Wärmedämmschicht –Systeme für Gasturbinen", PhD-Thesis Technical University Berlin, in German (1997).

[3]U. Schulz, K. Fritscher, C. Leyens, M. Peters, "Influence of processing on microstructure and performance of electron beam physical vapor deposition (EB-PVD) thermal barrier coatings", Jour. Engineering for Gas Turbines and Power, 124, 229-234 (2002).

[4]X. Zhao, P. Xiao, "Residual stresses in thermal barrier coatings measured by photoluminescence piezospectroscopy and indentation technique", Surf. & Coat. Technology 201, 1124-1131 (2006).

[5]M. R. Locatelli, E. R. Fuller, Jr., "Using OOF to model mechanical behaviour of thermal barrier coatings", presentation given at the int. conf. on adv. ceramics and composites, Cocoa Beach, FL (2002). http://www.ctcms.nist.gov/~fuller/PRESENTATIONS/index.html

[6]X. Zhao, X. Wang, P. Xiao, "Sintering and failure behaviour of EB-PVD thermal barrier coating after isothermal treatment", Surf. & Coat. Technology 200, 5946-5955 (2005).

[7]C. Leyens, U. Schulz, B.A. Pint, I.G. Wright, "Influence of electron beam physical vapor deposited thermal barrier coating microstructure on thermal barrier coating system performance under cyclic oxidation conditions", Surf. & Coat. Techn. 120-121, 68-76 (1999).

[8]T. Lauwagie, K. Lambrinou, I. Mircea, M. Bartsch, W. Heylen, O. Van der Biest, "Determining the elastic moduli of the individual component layers of cylindrical thermal barrier coatings by means of a mixed numerical-experimental technique", Materials Science Forum 492-493, 653-658 (2005).

[9]M. Bartsch, U.Schulz, B. Saruhan, "EB-PVD thermal barrier coatings for gas turbines, part 1: processing", Proceedings of the summer school of the European Research Training Network SICMAC – HRPN-CT-2002-00203, 11th-16th June 2006, Maó, Spain, 191-200.

THERMAL AND MECHANICAL PROPERTIES OF ZIRCONIA/MONAZITE-TYPE LaPO$_4$
NANOCOMPOSITES FABRICATED BY PECS

Seung-Ho Kim, Tohru Sekino, and Takafumi Kusunose
The Institute of Scientific and Industrial Research, Osaka University
8-1 Mihogaoka, Ibaraki, Osaka 567-0047, Japan

Ari T. Hirvonen
Materials Science and Engineering, Helsinki University of Technology
P.O. Box 6200, FI-02015 TKK, Finland

ABSTRACT
        Thermal barrier coatings (TBC's) perform the important function of insulating components such as gas turbine parts that operate at elevated temperatures. The most commonly applied TBC material is yttria stabilized zirconia (3YSZ), because it has a coefficient of thermal expansion similar to that of substrate metals. In this study, 3YSZ/monazite-type LaPO$_4$ nanocomposites were prepared by the pulse electric current sintering (PECS) method. The amount of LaPO$_4$ added to 3YSZ was varied from 0 to 40 vol.% and thermal and mechanical properties of these nanocomposites were investigated. The XRD results of the 3YSZ/LaPO$_4$ nanocomposites demonstrated differences in the crystalline phase of as-sintered and annealed zirconia. The phase transformation of as-sintered specimens was related to the amount of monazite-type LaPO$_4$, but this phenomenon was not observed in annealed specimens. The density of 3YSZ/LaPO$_4$ nanocomposites after annealing was decreased and the porosity was increased. Also, mechanical properties of 3YSZ/LaPO$_4$ nanocomposites decreased with increasing dispersion of monazite-type LaPO$_4$ particles. It was caused by low mechanical properties of LaPO$_4$ and weak bonding between 3YSZ. Thermal conductivity of 3YSZ/LaPO$_4$ nanocomposites was lower than 3YSZ. The difference of thermal conductivity between 3YSZ and 3YSZ/LaPO$_4$ nanocomposites at high temperatures was higher than that at low temperatures.

INTRODUCTION
        The continual development of high-tech industries, such as aerospace engineering and power plants etc., has required the development of new materials to replace traditional turbine materials, which have reached the limits of their temperature capabilities[1,2]. Thermal barrier coating (TBC) was proposed to improve the thermal efficiency of gas turbines due to the protective effect of the metal substrate. The protection materials of hot-path parts of gas turbine engines or jet engines were mainly developed using ceramics which have excellent thermal properties, chemical stability and heat insulation properties, etc[1,3,4]. The efficiency of gas turbines using ceramic TBC can be increased by 5 - 8 %. A ceramic coating of 300 μm thickness yields a temperature difference of up to 200 °C between the top coat and the metallic substrate.
        The selection of materials has been very important for the development of TBC. Some basic requirements of TBC materials is as follows: high melting point, no phase transformation for a variation of temperatures, low thermal conductivity, chemical stability, low mismatch of thermal expansion with the metallic substrate, good adherence to the metallic substrate and low sintering rate of the porous microstructure[1,5,6]. Specially, TBC materials must have low thermal conductivity and a high thermal expansion coefficient. TBC layers in the gas turbine should be considered for thermal stress, thermodynamic affinity between the bond- and top-coat, and

protection materials at high temperature, etc. The materials that can be applied to TBC have yttria stabilized zirconia (YSZ), a mixture of rare earth oxide, La$_2$Zr$_2$O$_7$, and zirconium phosphate, etc. In particular, the most commonly applied TBC material is YSZ because these materials have a coefficient of thermal expansion similar to that of the metallic substrate. The operation temperature of YSZ was limited below 1200 °C, because it undergoes a volume change as a result of a phase transformation from a monoclinic to tetragonal form at this temperature. To overcome this problem of YSZ, many researchers have studied how to improve its thermal properties such as high temperature stability, low thermal conductivity and high thermal expansion coefficient.

In this study, to improve the high temperature stability and low thermal conductivity of YSZ, monazite-type LaPO$_4$ was composed with YSZ to add properties such as weak bonding between oxides. 3YSZ/monazite-type LaPO$_4$ nanocomposites were fabricated by a pulse electric current sintering (PECS) method using composite powders, which consisted of chemically precipitated LaPO$_4$ on zirconia powder surfaces. The amount of monazite-type LaPO$_4$ added to 3YSZ varied from 0 to 40 vol.% and thermal and mechanical properties of 3YSZ/monazite-type LaPO$_4$ nanocomposites were investigated.

EXPERIMENTAL PRECEDURES

In this study, the composite powders were prepared by chemical precipitation[7,8] of monazite-type LaPO$_4$ on zirconia powder surfaces. The chemical precipitation of monazite-type LaPO$_4$ on zirconia powders was performed as follows. The raw material to make the Monazite-type LaPO$_4$ powder was La$_2$O$_3$ powder (99.9%, Shin-Etsu Chem. Co. Ltd., Japan). La$_2$O$_3$ powder was completely dissolved in HCl (6mol/l, Wako Pure Chem. Ind. Ltd., Japan). 3YSZ (TZ-3YE, Tosoh Corp., Japan) was ball-milled in a LaCl$_2$ solution for 12 h. The aqueous solution of H$_3$PO$_4$ (Wako Pure Chem. Ind. Ltd., Japan) was added on the slurry when the molar ratio of La to P was 1:1, and then the slurry was further mixed for 6 h. To precipitate monazite, the slurry was added to ammonia water (28%, Ishizu Seiyaku Ltd., Japan). This slurry was ball-milled for 6 h in order to homogenously distribute precipitation materials. After ball milling, the slurry was washed several times with de-ionized water and acetone. The washed powder was dried for 24 h in a dry oven at 55 °C. Dry powders were calcined at 700 °C for 2 h in atmospheric conditions. Calcined powders were ball-milled in a pot for 24 h. The specimens were fabricated by PECSed methods at 1300 °C, 30 MPa for 5 min in an Ar gas flow environment. The elevated temperature was 100 °C/min. The sintered body of composites was analyzed for thermal conductivity and its microstructure observed.

Density of the PECSed samples was measured by the Archimedes immersion method in toluene. Hardness and fracture toughness of sintered samples was measured by an indentation method. The indentation test of specimens was carried out by using a Vickers hardness tester (AVK-C2, Akashi Co. Ltd., Japan). Test conditions were made at the applied load of 98 N and the duration time of 15 seconds. The value of fracture toughness was calculated from the Niihara equation[9]. Flexural strength specimens of 3 x 4 x 36 mm in dimension were cut and ground from the sintered body, and the three point bending test was performed with conditions for crosshead speed of 0.5 mm/min with spans of 30 mm.

To confirm the microstructure of specimens, scanning electron microscopy (SEM, model S-5000, Hitachi Co. Ltd., Japan) was used. Specimens were fixed on the holder for SEM and coated with a thin evaporated gold film to avoid charging under the electron beam. The thermal conductivity (κ) was calculated from the equation,

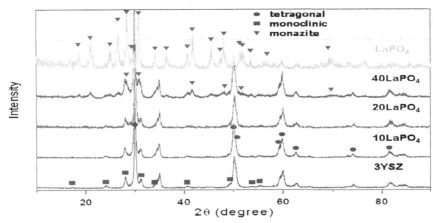

Figure 1. XRD patterns of 3YSZ/LaPO₄ compound powders.

$$\kappa = C_p \lambda \rho \qquad (1)$$

where C_p is the specific heat, λ is the thermal diffusivity and ρ is density. Bulk density of the sintering specimens was measured by the Archimedes immersion method using toluene solvent at room temperature. and the relative density was calculated with the theoretical density of each powder by the mixing rule. The thermal diffusivity was measured using the laser flash thermal constant analyzer (TC-7000, Ulvac-Riko, Japan). The specific heat was measured using the differential scanning calorimeter (DSC404C, NETZSCH, Germany).

RESULTS AND DISCUSSION

The XRD results of 3YSZ/LaPO₄ compound powder are shown in figure 1. 3YSZ had a mixture of the peak of tetragonal- and monoclinic-phases. Chemically precipitated LaPO₄ powder had a peak similar to that of the monazite-type. Figure 2 shows the XRD results of as-sintered and annealed 3YSZ/LaPO₄ nanocomposites. The XRD results of 3YSZ/LaPO₄ nanocomposites were different before (figure 2(a)) and after annealing (figure 2(b)) at 1500 °C. When LaPO₄ was added up to 30 vol.%, the XRD results of as-sintered nanocomposites showed an increase in the monoclinic phase, but that of nanocomposites with 40 vol.% LaPO₄ only showed a tetragonal-phase. The zirconia phase was affected by the amount of LaPO₄. When sintered nanocomposites were annealed at 1500 °C for 1 h, the XRD results of the annealed nanocomposites were different as compared to that of sintered nanocomposites. The XRD results of annealed nanocomposites were similar to all compositions. The XRD results of annealed nanocomposites showed a peak of a mixture of tetragonal- and monoclinic-phases, as shown in figure 2(b). The peaks after annealing were sharper than as-sintered specimens. The phase transformation of zirconia to the amount of LaPO₄ was not observed in the annealed specimen.

Figure 3 shows the density and porosity of 3YSZ/LaPO₄ nanocomposites before and after annealing. Density and porosity was measured by the Archimedes immersion method using toluene solvent at room temperature. As the 3YSZ/LaPO₄ nanocomposites annealed, the

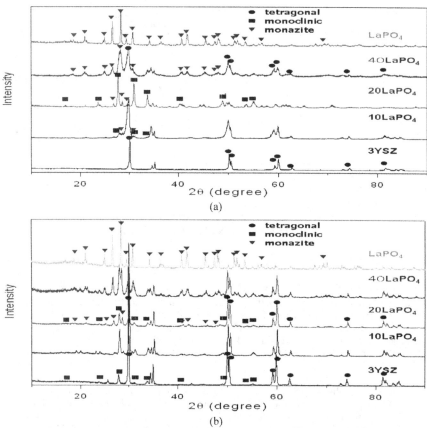

Figure 2. XRD patterns of 3YSZ/LaPO$_4$ nanocomposites (a) before- (as-sintered) and (b) after annealing at 1500 $^{\circ}$C for 1 h in atmosphere.

experimental density decreased, but porosity increased. However, the density and porosity of monolithic LaPO$_4$ was increased. The decreasing density of 3YSZ/LaPO$_4$ nanocomposites after annealing was related to the phase transformation shown in figure 2. It suggested that the decreasing density of 3YSZ/LaPO$_4$ nanocomposites was caused by an increasing porosity as a result of a volume change resulting from a phase transformation from a monoclinic- to tetragonal-phase. In annealed 3YSZ/LaPO$_4$ nanocomposites, the relative density decreased from 97.9 – 99.3 % of theoretical density to 95.3 – 96.6 %. On the other hand, the density of monolithic LaPO$_4$ increased from 97.0 to 99.6 %. Figure 4 shows the mechanical properties of 3YSZ/LaPO$_4$ nanocomposites as a function of the amounts of monazite-type LaPO$_4$ particles. As the amount of monazite-type LaPO$_4$ particles increased, mechanical properties of 3YSZ/LaPO$_4$

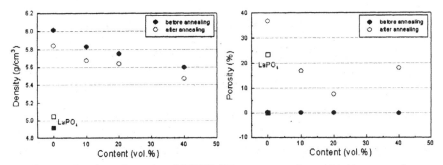

Figure 3. Density and porosity of 3YSZ/LaPO$_4$ nanocomposites with/without annealing at 1500 °C for 1 h in atmosphere.

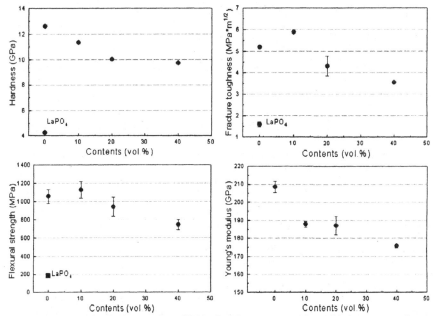

Figure 4. Mechanical properties of 3YSZ/LaPO$_4$ nanocomposites with the amount of monazite-type LaPO$_4$.

nanocomposites decreased. The degradation of mechanical properties was caused by dispersed monazite-type LaPO$_4$ particles. Monazite-type LaPO$_4$ materials have poor mechanical properties and weak bonding between oxides[10]. Actually, mechanical properties of PECSed monolithic

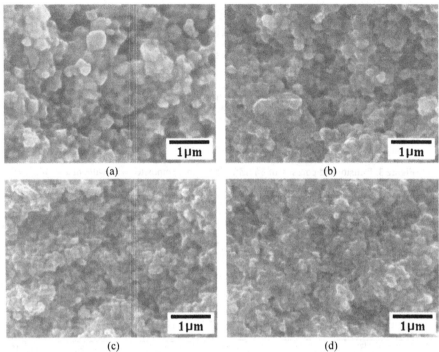

Figure 5. SEM micrographs of the fracture surfaces of 3YSZ/LaPO₄ nanocomposites with the varying amounts of monazite-type LaPO₄; (a) 3YSZ, (b) 10 vol.% LaPO₄, (c) 20 vol.% LaPO₄ and (d) 40 vol.% LaPO₄.

LaPO₄ have a hardness of 4.26 GPa, 1.60 MPa·m$^{1/2}$ of fracture toughness and 174.9 MPa of flexural strength. Young's modulus can not be measured by range error. Mechanical properties of 3YSZ/LaPO₄ nanocomposites were decreased due to dispersed monazite-type LaPO₄ particles. Mechanical properties of 3YSZ/LaPO₄ nanocomposites were lower than 3YSZ. In 3YSZ/10 vol.% LaPO₄ nanocomposites, flexural strength was higher (~1127 MPa) than other nanocomposites. When 40 vol.% LaPO₄ particles was added, flexural strength of 3YSZ/LaPO₄ nanocomposites increased to more than 700 MPa.

The fracture behaviors of 3YSZ and 3YSZ/LaPO₄ nanocomposites were compared as shown in figure 5. As the amount of monazite-type LaPO₄ increased, the grain size of 3YSZ/LaPO₄ nanocomposites gradually decreased. The grain size of 3YSZ and 3YSZ/40 vol.% LaPO₄ nanocomposites were about 200-300 nm and about 100-150 nm, respectively. This effect was likely due to the addition of dispersed monazite-type LaPO₄ particles, which likely suppressed grain growth in 3YSZ. In general, decreasing grain size increases fracture strength due to a Hall-Petch relationship. However, in these systems, the flexural strength of 3YSZ/LaPO₄ nanocomposites by dispersed monazite-type LaPO₄ decreased in spite of

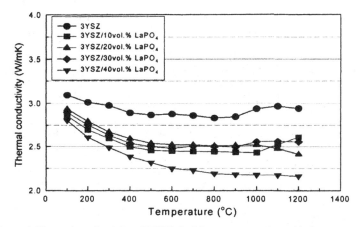

Figure 6. Thermal conductivity of 3YSZ/ LaPO₄ nanocomposites with the amount of monazite-type LaPO₄ and temperature.

decreasing grain size. The degradation of flexural strength was due to weak bonding between 3YSZ and LaPO₄. For this reason, fracture behaviors were dominated by intergranular-type fracture along the grain boundaries, as shown in figure 5.

Figure 6 shows the thermal conductivity of PECSed 3YSZ/monazite-type LaPO₄ nanocomposites with monazite-type LaPO₄ contents. Thermal conductivity of 3YSZ/monazite-type LaPO₄ nanocomposites was lower than the 3YSZ monolith. As the amount of monazite-type LaPO₄ increased, thermal conductivity of 3YSZ/LaPO₄ nanocomposites decreased. Also, the stabilization range of thermal conductivity of specimens was wider than monolithic 3YSZ. Thermal conductivity of monolithic 3YSZ slightly decreased or was constant up to 900 °C, and increased above 900 °C. In the case of nanocomposites, these phenomena were observed at higher temperatures than the monolith. The temperature of nanocomposites observed in these phenomena was above 1000 °C or more. These phenomena were related to grain size and the amount of LaPO₄. As the amount of LaPO₄ increased, weak boundaries between 3YSZ and LaPO₄ increased. Thermal conductivity was related to phonon diffraction generated by grain boundaries such as 3YSZ/3YSZ and 3YSZ/LaPO₄. The effect of phonon diffraction in weak boundaries was higher than that of strong interfaces. Therefore, thermal conductivity was likely caused by increasing phonon diffraction due to increments in areas of grain boundaries. The stable range of nanocomposites was longer than the monolith, as shown in figure 6. The difference of thermal conductivity between 3YSZ and 3YSZ/LaPO₄ nanocomposites increased with elevating temperature and the amount of LaPO₄. In 3YSZ/40 vol.% LaPO₄ nanocomposites, thermal conductivity was reduced to 9 % at 100 °C and 27 % at 1200 °C , respectively, when compared with 3YSZ. This indicated that the high temperature stability of monazite-type LaPO₄ added to 3YSZ improved as compared to the monolith.

CONCLUSIONS

In order for compounds to be considered as TBC materials, they must have high temperature stability and low thermal conductivity, etc., In order to improve high temperature stability and low thermal conductivity of YSZ, monazite-type $LaPO_4$ was composed with 3YSZ to add properties such as weak bonding between zirconia. Monazite-type $LaPO_4$ was prepared by chemical precipitation methods. The XRD results of 3YSZ/LaPO₄ nanocomposites demonstrated differences in the crystalline phases between as-sintered and annealed zirconia. The phase transformation of as-sintered specimens was related to the amount of monazite-type $LaPO_4$, but annealed specimens were not observed to be affected by the amount of monazite-type $LaPO_4$ on phase transformation. The density of 3YSZ/LaPO₄ nanocomposites after annealing was decreased and porosity was increased. In addition, mechanical properties of 3YSZ/LaPO₄ nanocomposites decreased with increasing dispersion of monazite-type $LaPO_4$ particles. This was likely caused by low mechanical properties of $LaPO_4$ and weak bonding between 3YSZ. Thermal conductivity of 3YSZ/LaPO₄ nanocomposites was lower than 3YSZ. The difference in thermal conductivity between 3YSZ and 3YSZ/LaPO₄ nanocomposites at high temperatures was higher than that at low temperatures.

ACKNOWLEDGMENTS

This work was carried out as a part of the "Nanostructure Coating Project (Nano-Coating Functions, Structural Design and Control Technique)" under the Nanotechnology Materials Program supported by The New Energy and Industrial Technology Development Organization (NEDO), Japan.

REFERENCES

[1]X.Q. Cao, R. Vassen, and D. Stoever, "Ceramic Materials for Thermal Barrier Coatings," *J. Euro. Ceram. Soc.*, **24**, 1-10 (2004).

[2]D.R. Clarke, and S.R. Phillpot, "Thermal Barrier Coating Materials," *Materialstoday*, 22-29 (2005).

[3]U. Schulz, B. Saruhan, K. Fritscher, and C. Leyens, "Review on Advanced EB-PVD Ceramic Topcoats for TBC Applications," *Int. J. Appl. Ceram. Technol.*, **1**, 302-15 (2004).

[4]D. Zhu, and R.A. Miller, "Thermal Barrier Coatings for Advanced Gas Turbine and Diesel Engines," *NASA Glenn Research Center, Cleveland, Ohio, NASA/TM-209453*, 1-12 (1999).

[5]X. Cao, R.Vassen, W. Fischer, F. Tietz, W. Jungen, and Detlev Stöver, "Lanthanum-Cerium Oxide as a Thermal Barrier-Coating Material for High-Temperature Applications," *Adv. Mater.*, **15**, 1438-42 (2003).

[6]F. Cernuschi, P. Bianchi, M. Leoni, and P. Scardi, "Thermal Diffusivity/Microstructure Relationship in Y-PSZ Thermal Barrier Coatings," *J. Therm. Spray Technol.*, **8**, 102-109 (1999).

[7]W. Min, K. Daimon, T. Matsubara, and Y. Hikichi, "Thermal and Mechanical Properties of Sintered Machinable LaPO₄-ZrO₂ Composites," *Mater. Res. Bull.*, **37**, 1107-15 (2002)

[8]W. Min, D. Miyahara, K. Yokoi, T. Yamaguchi, K. Daimon, Y. Hikichi, T. Matsubara, and T. Ota, "Thermal and Mechanical Properties of Sintered LaPO₄-Al₂O₃ Composites," *Mater. Res. Bull.*, **36**, 936-45 (2001)

[9]K. Niihara, "Indentation Microfracture of Ceramics - Its Application and Problems," *Ceramics Japan*, **20**, 12-18 (1985).

[10]J.B. Davis, D.B. Marshall, R.M. Housley, and P.E.D. Morgan, "Machinable Ceramics Containing Rare-Earth Phosphates," *J. Am. Ceram. Soc.*, **81**, 2169-75 (1998).

CORROSION BEHAVIOR OF NEW THERMAL BARRIER COATINGS

R. Vaßen, D. Sebold, D. Stöver
Institut für Energieforschung (IEF-1)
Forschungszentrum Jülich GmbH
Jülich, 52425, Germany

ABSTRACT
    New thermal barrier coating (TBC) systems based on pyrochlores ($La_2Zr_2O_7$, $Gd_2Zr_2O_7$, $La_2Hf_2O_7$) have been produced by atmospheric plasma-spraying. Double layer systems consisting of a 200 μm yttria stabilized zirconia (YSZ) layer on the bondcoat and a 200 μm pyrochlore layer on top have been used. These coatings have been tested in a gas burner test rig. In the flame of the rig corrosive media have been injected. These media were Na_2SO_4 and NaCl as a water solution and also kerosene. The lifetime of the coatings was largely reduced by the injection of the corrosive species. A detailed analysis of the failure mechanisms and a comparison to YSZ coatings will be made.

INTRODUCTION

Thermal barrier coatings systems typically consist of a metallic oxidation protection layer and an insulative ceramic topcoat. Electron beam physical vapour deposition (EB-PVD) and atmospheric plasma spraying (APS) are widely used processes to deposit the topcoat of these systems. The state of the art topcoat material for both processes is yttria partially stabilized zirconia (YSZ) [1, 2]. This material performs well up to about 1200 °C. At higher application temperatures, which are envisaged for a further improvement of the efficiency of the gas turbines, the YSZ undergoes two detrimental changes. Significant sintering leads to microstructural changes and hence a reduction of the strain tolerance in combination with an increase of the Young's modulus. Higher stresses will originate in the coating, which lead to a reduced life under thermal cyclic loading.

The second change is a phase change of the non-transformable t´-phase, which is present in the as-deposited YSZ coating. At elevated temperatures the t´-phase transforms into tetragonal and cubic phase. During cooling the tetragonal phase will further transform into the monoclinic phase, which is accompanied by a volume change and a high risk for a damage of the coating [3]. As a consequence, a considerable reduction of thermal cycling life is observed.

These disadvantageous properties of YSZ at high temperatures prompted an intense search for new TBC materials in the past. In [2, 4, 5] detailed overviews on the developments of new systems are given.

Among the interesting candidates for thermal barrier coatings, those materials with pyrochlore structures and high melting points show promising thermo-physical properties. Interesting candidates are especially $La_2Zr_2O_7$ $La_2Hf_2O_7$, $Gd_2Zr_2O_7$ or $Nd_2Zr_2O_7$. Previous investigations show excellent physical properties of theses materials, i.e. thermal conductivity lower than YSZ and high thermal stability [6]. However, relatively low thermal expansion coefficients and toughness values are observed in these materials [7]. As a result, the thermal cycling properties are worse than those of YSZ coatings. A way to overcome this shortcoming is the use of layered topcoats. The failure of TBC systems often occur within the TBC close to the bondcoat/topcoat interface. At this location YSZ is used as a TBC material with a relatively high thermal expansion coefficient and high toughness. The YSZ layer is then coated with the new TBC material (e.g. $La_2Zr_2O_7$) which is able to withstand the typically higher temperatures at this location. In the past years in several publications on $YSZ/La_2Zr_2O_7$ double layer systems we show that this concept really works [8, 9, 10, 11].

This study is now focused on the influence of corrosive media on the performance of new TBC systems. Corrosive media can be introduced from the environment e.g. in aviation engines from sand particles during landing and take-off. In addition, also the fuel contains corrosive constituents, the most harmful are vanadium, sulphur and sodium. The concentration can vary in wide ranges. While for aviation fuel the sulphur content is limited to 0.05 wt.%, it may reach 4 wt.% in heavy oil for stationary gas turbines [12].

A large number of investigations have been performed in the past to study the influence of the corrosive media in laboratory tests. An older review of the results has been made by Bürgel and Kvernes [13] and a more recent one by Jones [14], showing the most important reactions of the corrosive media with the TBCs. Under severe corrosive environments it is found that the stabilizing agent reacts with the corrosive media leading to a destabilizing of the coating. Also new stabilizing agents as Sc_2O_3 have been studied [15]. A distinct improvement of the corrosion behaviour of these materials compared to YSZ was not found.

Only few investigations on new TBC materials have been published. In [16] $La_2Zr_2O_7$ coatings have been compared to YSZ coatings under isothermal conditions. The new TBC material was relatively resistant against vanadia attack at 1000°C, while a fast decomposition in sulphur containing environment at 900°C was observed.

Corrosion testing under isothermal conditions in a furnace or in burner rigs testing [17, 18] show a major influence of the bond coat composition on the performance of the TBC systems. Typically higher Cr and Al contents improve the lifetime of the coatings. These results show that the corrosion of the bond coat and the attack of the formed thermally grown oxide (typically an alumina scale) play a major role. In the temperature range above 800°C so-called type I hot corrosion is expected. Reactions of the formed Al_2O_3 scale with SO_3 containing atmospheres can be as follows [19], with (1) being dominant at high temperatures (type I hot corrosion):

$$Al_2O_3 + O^{2-} \rightarrow 2\,AlO_2^- \quad \text{for low } SO_3 \text{ pressures} \tag{1}$$

$$Al_2O_3 + 3SO_3 \rightarrow Al_2(SO_4)_3 \quad \text{for high } SO_3 \text{ pressures} \tag{2}$$

In addition, under type I conditions, the sulphide formation of metals (Cr, Ni) at the slag/metal interface play an important role:

$$M + SO_2 \rightarrow M\text{-oxide} + M\text{-sulphide} \tag{3}$$

In the present investigation corrosive media are injected into the flame of a burner rig. With the used setup temperature profiles can be adjusted which are considered as relevant for modern gas turbines (i.e. bond coat temperatures in the range of 900 to 1000°C and surface temperature of about 1200°). Hence a rather realistic testing of TBC systems under corrosive conditions should be possible. In a previous paper already the results of the tests on YSZ coatings have been published [20].

EXPERIMENTAL

The investigated thermal barrier coating systems have been produced by plasma spraying with two Sulzer Metco plasma-spray units. Vacuum plasma spraying with a F4 gun was used to deposit a 150 µm NiCo21Cr17Al13Y0.6 bond coat (Ni 192-8 powder by Praxair Surface Technologies Inc., Indianapolis, IN) on disk shaped IN738 superalloy substrates. The diameters of the substrates used for thermal cycling tests were 30 mm, the thickness 3 mm. At the outer edge a radius of curvature of 1.5 mm was machined to avoid sharp edges.

The ceramic top coats with a thickness of about 400 µm were produced by atmospheric plasma spraying (APS) using a Triplex I gun. For the new TBCs a double layer structure was used with an about 200 µm YSZ layer on the bondcoat and an about 200 µm coating of the new TBC material on top. During the manufacture of the thermal cycling specimens also steel substrates were coated. These coatings were used to characterize the as-sprayed condition. The Argon and Helium plasma gas flow rates were 20 and 13 standard liter per minute (slpm), the plasma current was 300 A at a power of 20 kW.

Corrosion tests were performed in a gas burner rig setup, in which the disk shaped specimens were periodically heated up to the desired surface temperature of about 1200 °C in approximately 1 min by a natural gas/ oxygen burner. After 5 min heating the specimens were cooled for 2 min from both sides of the specimens by compressed air. The surface temperature was measured with an infrared pyrometer operating at a wavelength of 9.6 - 11.5 µm and a spot size of 5 mm.The substrate temperature was measured using a NiCr / Ni thermocouple inside a hole drilled to the middle of the substrate. All given values of temperatures are mean values taken over all cycles after completion of the heating phase. More details on the standard thermal cycling rigs can be found in [21]. In this investigation the central gas nozzle was used to introduce atomized liquids containing the corrosive media into the flame. Results on water based salt solutions and on kerosene will be presented.

The injection of water droplets into the flame led to a reduced stability of the flame and therefore larger variations in the temperature profile. In order to evaluate this effect also a pure water injection (without additional corrosive media) was examined for YSZ coatings. Here no significant influence of the water injection on the thermal cycling performance was found.

During the testing 1.4 g/min (salt solution) or 1.2 g/min (kerosene) of corrosive containing media has been injected into the flame. The corrosive media consisted of distilled water with 1 wt.-% Na_2SO_4, 1 or 5 wt.-% NaCl, or of kerosene. The total mean methane and oxygen gas flows during thermal cycling was about 550 l/min and 670 l/min giving about 0.22 wt.-% of corrosive salt species to natural gas for the 1 % solutions.

It turned out that the gas flows of the burner gases varied by about 25 % to adjust similar temperature profiles. This fact led to the rather large differences in the corrosive media concentration between 0.17 to 0.25 wt.-%. As the injection nozzle is located in the center of the flame the profile of the corrosive media concentration has a maximum in the centre.

The test was stopped when obvious degradation of the coatings (large delamination) occurred during the thermal cycling. This definition of failure results in an uncertainty in the lifetime data especially if partial delamination of the coatings occurs. However, as seen below the resulting error probably does not effect the interpretation of the results.

Sprayed specimens were vacuum impregnated with epoxy, and then sectioned, ground and polished. As for the preparation water was used as a media, water-soluble corrosion products might be removed by the preparation process. Cross-sections of the coatings were examined by optical microscopy and scanning electron microscopy (Ultra 55, Zeiss). The surface of some of the tested samples were analysed by X-ray diffraction using a Siemens D5000 facility at a wavelength of 1.5406 Å. For some samples also a special preparation technique was used. About 2 mm thick stripes were prepared from the center of the thermal cycling samples by laser cutting. In the center of these stripes an additional cut was made through the metal substrate. The ceramic coating was broken along this cut and the fracture surface investigated. This was made to check whether the water based preparation methods lead to a removal or redistribution of the corrosive species within the sample. It turned out that also on the fracture surface corrosive species were clearly found similar to the results of the polished samples. However, the quality of the images was purer than for the cross-sections. Therefore only results of the cross-sections will be presented here.

RESULTS AND DISCUSSION

The microstructure of the new TBCs and the YSZ coatings are shown in Fig. 1. The porosity level of the coatings measured by mercury porosimetry was about 12 vol.-%.

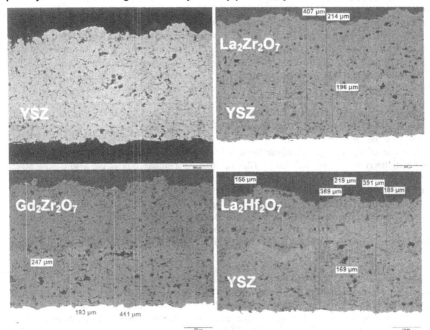

Fig. 1 Micrographs of the investigated TBCs.

Fig. 2 shows the performance of YSZ TBCs in a cyclic burner rig without the addition of corrosive species. The lifetime of the systems under the conditions used in this study ($T_{surface}$ between 1100 and 1250°C) is well above 1500 cycles. An investigation of the influence of the injection of pure water has been performed for YSZ coatings. A lifetime of more than 4000 cycles was found which is comparable to the lifetime of YSZ coatings without water injection for the given bondcoat and surface temperatures.

The results of the cyclic testing of the new double layers and YSZ coatings are also summarized in Fig. 2. It is obvious that the lifetime of single layer $La_2Zr_2O_7$ coatings is reduced compared to our standard YSZ coatings. The YSZ coatings show a strong reduction of lifetime at surface temperatures above about 1300 °C. Double layer systems made of YSZ and $La_2Zr_2O_7$ perform excellent even at much higher temperatures. Their temperature capability is under the given cyclic conditions in the range of 1450°C. Also the one tested YSZ/$Gd_2Zr_2O_7$ double layer system performed well. In contrast, the YSZ/$La_2Hf_2O_7$ coating showed an early failure with a spallation of individual spray splats from the surface. This result indicates a lower temperature capability of the hafnate compared to the zirconate materials.

At lower surface temperatures (< 1300°C) the double layer systems made of YSZ/$La_2Zr_2O_7$ perform similar as the YSZ coatings. Failure is then related to crack growth within the ceramic close to the bondcoat which is induced by the growth of the so-called thermally grown oxide (TGO) on top of the bondcoat. Here for both types of coatings a YSZ layer is present leading to similar performance data.

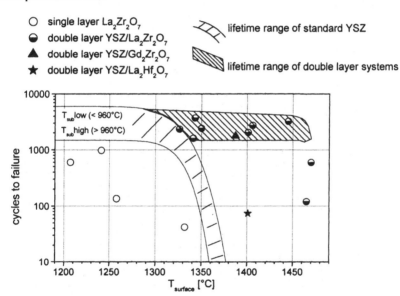

Fig. 2 Cycles to failure for single and double layer systems in a cyclic burner rig without addition of corrosive species as a function of the surface temperature.

In Fig. 3 the results of the cyclic testing with corrosive media are summarized. More details especially on the mean temperature during testing are given in Table I. While the substrate temperatures were in all cases between 890 and 950°C except for 5% NaCl, the surface temperature showed some larger variations. Typically values between 1150 and 1200°C could be established for the 1% salt and the kerosene addition. For the higher salt loading (5%) it was not possible to obtain sufficiently high surface temperatures. Also the substrate temperatures were reduced. The reason is possibly the melting of the NaCl on the surface of the samples. The heat of fusion of NaCl (melting temperature 801°C), leads to a heat consumption for the 5% media in an one cm² large area in the center of the sample of nearly 100kW/m², hence effectively cooling the surface.

A comparison with Fig. 1 shows that all samples failed earlier under the influence of the corrosive media than without. At the given substrate temperatures well below 950°C lifetimes of several thousand cycles are expected.

The highest reduction of more than a factor of 100 is observed for all samples for the media with the high concentration of NaCl (5 %). Compared to this result the reduced concentration (1 %) led to clearly increased lifetimes. Samples cycled with Na₂SO₄ media show lifetimes in between those of samples cycled with high and low NaCl concentration. The addition of kerosene had a reduced effect on the lifetime. Only for the YSZ/GZ coatings a pronounced reduction was observed.

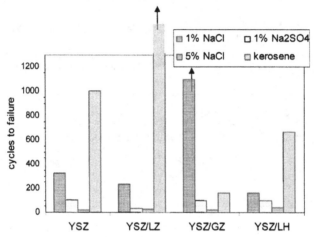

Fig. 3 Cycles to failure of YSZ and double layers of YSZ/La₂Zr₂O₇ (YSZ/LZ), YSZ/La₂Hf₂O₇ (YSZ/LH), and YSZ/Gd₂Zr₂O₇ (YSZ/GZ) in a burner rig test with addition of NaCl (1 % or 5 %, 1.4 g/min), Na₂SO₄ (1%, 1.4 g/min), and kerosene (1.2 g/min). The YSZ/GZ did not fail after 1100 cycles with 1 % NaCl addition. The lifetime of the YSZ/LZ sample with kerosene addition was 1750 cycles.

Photos of the thermally cycled samples are found in Fig. 4. The types of failure can be roughly divided into two groups. The first type is characterized by a spallation of the whole coating, correspondingly the bondcoat is visible on the photos in Fig. 2. In the second type of failure the ceramic coating itself is heavily attacked by the corrosive media leading to a partly delamination of the coating. With the exception of the YSZ/GZ system the 1% NaCl and the Na₂SO₄ addition led to failure mode one although some coloring of the coatings is observed, while 5% NaCl seem to promote the corrosive attack of the ceramic itself. The samples cycled with kerosene addition show some mixed kind of failure.

The failure mechanisms of the coatings cycled under the corrosive environment will be discussed in the following. It will also be investigated whether in failure mode 1 already an attack of the ceramic coating took place.

Table I. Summary of experimental details of the corrosion rig testing as mean substrate and surface temperature (T_{sub}, T_{surf}), cycles to failure, sample number (#), * indicates that the sample was not yet damaged.

	YSZ	YSZ/LZ	YSZ/GZ	YSZ/LH
corrosion media	1% NaCl			
#	884	860	872	879
Tsurface [°C]	1156	1182	1242	1215
Tsub[°C]	902	913	934	918
cycles to failure	325	232	1100*	159
corrosion media	1% Na2SO4			
#	885	861	873	880
Tsurface [°C]	1119	1163	1214	1142
Tsub[°C]	923	895	932	933
cycles to failure	102	32	98	97
corrosion media	5%NaCl			
#	887	863	875	872
Tsurface [°C]	1043	931	1004	1035
Tsub[°C]	843	858	804	913
cycles to failure	21	25	19	40
corrosion media	kerosene			
#	928	862	874	881
Tsurface [°C]	1167	1217	1184	1226
Tsub[°C]	947	931	940	933
cycles to failure	1004	1750	160	663

In Fig. 5 micrographs of some samples cycled with 1% NaCl and 1% Na$_2$SO$_4$ are shown. In most of the samples an increase of the porosity level is found in comparison to the as-sprayed condition (Fig. 1). This will be discussed later. In addition, also a dark layer of thermally grown oxide (TGO) is found on the bondcoat having a thickness between about 2 and 5 μm. This appears to be very thick for the low bondcoat temperature (< 980°C) and the short time at temperature (< 100 h). On the other hand the β-phase depleted zone in the top of the bondcoat is very thin. A depleted zone is typically seen in TBC bondcoats as the TGO is mainly Al$_2$O$_3$ which is formed from the Al of the β-phase (mainly NiAl). The reason for the thin depleted zone can be identified if the TGO is analyzed in more detail. Fig. 6 shows a SEM micrograph of a

$YSZ/La_2Zr_2O_7$ system and an EDX spectra of the indicated area after cycling with NaCl addition. Obviously, the oxide does not only consist of alumina, however especially on top more complex phases consisting of Ni, Co, Cr, Al and others are found. The corrosive species lead to a fluxing of the alumina scale forming a much thicker, complex oxide layer. This thick layer is expected to have low mechanical strength and introduces additional stresses into the coating system leading to an early failure of the TBC system. The failure of the investigated samples cycled with Na_2SO_4 and 1 NaCl addition is mainly the result of the TGO attack by the corrosive species. For the high NaCl concentration the time at high temperatures is too short to promote the corrosion of the TGO, for the kerosene addition the amount of corrosive species seems to be too low for a distinct attack.

1% NaCl

1% Na_2SO_4

5% NaCl

Kerosene

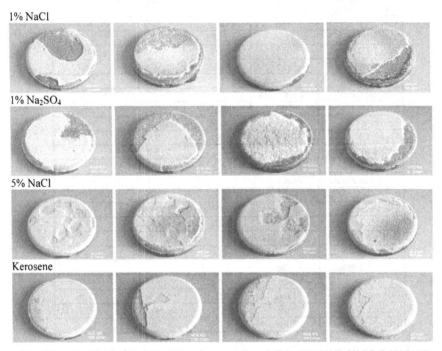

Fig. 4 Photos of the samples after corrosion tests, from left to right YSZ, YSZ/LZ, YSZ/GZ, YSZ/LH.

The attack of the ceramic layers will be discussed in the following. EDX analyis of the TBCs showed at many locations of the samples corrosive species containing sodium, sulfur, chloride but also chromium or nickel. These corrosive species could be especially detected in the fracture surfaces as there a removal of the water soluble species was avoided. These corrosive compounds lead to a stresses during thermal cycling and crack growth in the samples. Also a

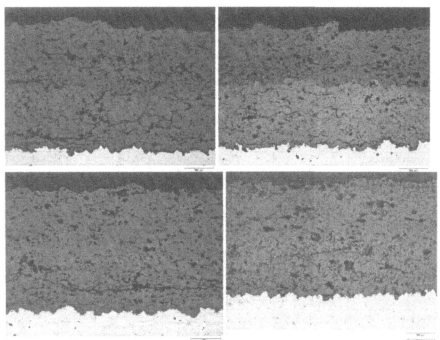

Fig. 5 Optical Micrographs of samples thermally cycled with 1% NaCl addition (top, left YSZ/LZ, right YSZ/LH) and Na$_2$SO$_4$ (bottom, left YSZ/LZ, right YSZ/LH).

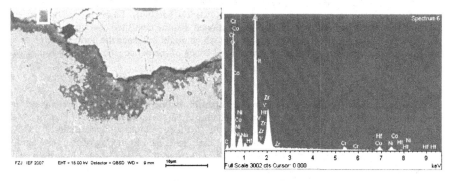

Fig. 6 SEM micrographs of a YSZ/La$_2$Zr$_2$O$_7$ sample after thermal cycling with NaCl addition (top, left). The other graph shows the EDX analysis of the indicated area.

reaction with some of the ceramics was observed leading for example to the formation of non-stabilized zirconia and also to the formation of higher porosity levels. It should be mentioned here that part of the corrosive species within the ceramic could also come from the substrate holder made of a Ni base alloy.

In the case of the samples cycled with 5% NaCl addition the filling of the cracks with high amounts of NaCl salt result in a very early failure of the samples. It is assumed that the strain tolerance of the samples is lost due to the filled cracks and pores.

Due to the high number of investigated systems not all the results can be presented here. Therefore it was tried to summarize the results of the cyclic tests in Table II.

Table II. Summary of the thermal cycling behaviour under corrosive conditions. The numbers indicate the degree of corrosive attack: 0 hardly any attack, 1 some corrosion or crack formation, 2 distinct corrosion or crack formation, 3 massive attack. X: results of X-ray phase analysis.

Corrosive attack		YSZ	YSZ/LZ	YSZ/GZ	YSZ/LH
1% NaCl	BC	2	2	2	2
	TBC	2 (cracks, X: monoclinic ZrO_2)	1-2 (cracks, X: no phase changes)	1 (some cracks, no failure!, long test time)	3 (cracks, X: new phases)
1% Na_2SO_4	BC	3	3	3	3
	TBC	2-3	1-2 (short test time)	3 (new phases?)	2
5% NaCl	BC	0	0	0	0
	TBC	3 (X: no phase changes)	2 (cracks, X:monoclinic ZrO_2)	2 (cracks, X: no phase changes)	2-3 (top layer removed)
Kerosene	BC	1	1	1	1
	TBC	1-2	2 (long test time)	1 (short test time)	2-3

Under the given conditions $La_2Zr_2O_7$ and $Gd_2Zr_2O_7$ show a equal or slightly better performance than YSZ (with the exception of $Gd_2Zr_2O_7$ with Na_2SO_4 addition). $La_2Hf_2O_7$ has a reduced corrosive stability, however it should also be mentioned that the deposited coating showed some loss of lanthanum due to the thermal spray process which might influence the performance.

Table II also clearly indicates the more pronounced attack of the bondcoat by Na_2SO_4 compared to NaCl.

CONCLUSIONS

The results of a new burner rig allowing corrosion testing of various TBC systems under rather realistic conditions have been presented. The injection of NaCl, Na_2SO_4 and kerosene led to a distinct reduction of lifetime up to a factor of 100 for the highest concentration (> 1%). Failure of the coatings was for the moderate concentrations related to fluxing of the TGO oxide by the corrosive species. Clearly, the Na_2SO_4 addition attacked the TGO faster than the NaCl addition.

Also in the case of a TGO fluxing an additional attack of the ceramics was observed. Compared to YSZ $La_2Zr_2O_7$ and $Gd_2Zr_2O_7$ performed rather good, $La_2Hf_2O_7$ showed a lower stability under corrosive conditions. Failure was induced by the filling of the cracks with corrosive species and in some cases by a reaction of the ceramics with the corrosive species.

The addition of the solution with high NaCl concentration led to early failure which was explained by a significant loss of the strain tolerance by the filling of the crack network with salt species.

ACKNOWLEDGEMENT
The authors would like to thank Mr. K.H. Rauwald and Mr. R. Laufs (both IEF-1, FZ Jülich) for the manufacture of the plasma-sprayed coatings and Mrs. N. Adels and Mrs. A. Hilgers (both IEF-1) for the thermal cycling of the specimens. The authors also gratefully acknowledge the work of Mrs. H. Moitroux (IEF-1), Mr. P. Lersch (IEF-2), Mrs. S. Schwartz-Lückge (IEF-1), Dr. D. Sebold (IEF-1) and Mr. M. Kappertz (IEF-1) who supported the characterization of the samples by photography, XRD, optical and scanning electron microscopy, and sample preparation.

REFERENCES
1 W.A Nelson,. R.M. Orenstein, TBC Experience in Land-Based Gas Turbines, *Journal of Thermal Spray Technology* **6**, 2 , 176-180, (1997).
2 D.R. Clarke and C.G. Levi, Annu. Rev. Mater. Res. **33**, 383-417, (2003).
3 R. A. Miller J.L. Smialek, R.G. Garlick, Phase Stability in Plasma-Sprayed Partially Stabilized Zirconia-Yttria, in Science and Technology of Zirconia, Advances in Ceramics, Vol. 3, A.H. Heuer and L.W. Hobbs (eds.), The American Ceramic Society, Columbus, OH, USA, 241-251, (1981).
4 R. Vaßen, D. Stöver, Conventional and new materials for thermal barrier coatings, in Functional Gradient Materials and Surface Layers Prepared by Fine Particle Technlogy, NATO Science Series II: Mathematics , Physics and Chemistry - Vol. 16, Kluwer Acadmic Publishers, Dordrecht, The Netherlands 199-216, (2001).
5 J.R. Nicholls, Advances in Coating Design for High-Performance Gas Turbines, *MRS Bulletin*, Sept., 659-670 (2003).
6 H. Lehmann, D. Pitzer, G. Pracht, R. Vaßen, D. Stöver, Thermal Conductivity and Thermal Expansion Coefficients of the Lanthanum-Rare Earth Element-Zirconate System, *J. Amer. Ceram. Soc.*, **86,** 8 1338-44, (2003).
7 U. Bast, E. Schumann, "Development of Novel Oxide Materials for TBCs, *Ceramics Engineering & Science Proceedings*, 23, 4 (2002) 525-32.
8 R. Vaßen, G. Pracht, D. Stöver, New Thermal Barrier Coating Systems with a Graded Ceramic Coating, Proc. of the International Thermal Spray Conference 2002, Verlag für Schweißen und verwandte Verfahren DVS-Verlag GmbH, Düsseldorf, 2001, pp. 202-207.
9 R. Vaßen, G. Barbezat, D.Stöver, Comparison of Thermal Cycling Life of YSZ and $La_2Zr_2O_7$-Based Thermal Barrier Coatings, in Materials for Advanced Power Engineering 2002, eds. J. Lecomte-Becker, M. Carton, F. Schubert, P.J. Ennis, Schriften des Forschungszentrum Jülich, Reihe Energietechnik, **21**, 1, 511-521.
10 R. Vaßen, X.Q. Cao, D. Stöver, Improvement of New Thermal Barrier Coating Systems using a Layered or Graded Structure, *Ceramic Engineering & Science Proceedings*, **22**, 4, 435- 442 (2001).
11 R. Vaßen, M. Dietrich, H. Lehmann, X. Cao, g. Pracht, F. Tietz, D. Pitzer, D. Stöver, Development of Oxide Ceramics for an Application as TBC", *Materialwissenschaft und Werkstofftechnik* **8**, 673-677, (2001).

12 B.R. Marple, J. Voyer, C. Moreau, D.R. Navy, Corrosion of Thermal Barrier Coatings by Vanadium and Sulfur Components, Materials at High Temperatures, 17, 3, 397-412 (2000).

13 R. Bürgel, I. Kvernes, Thermal Barrier Coatings, High Temperature Alloys for GasTurbines and Other Applications 1986, W. Betz et al. , ed. D. Reidel Publiching , 327-356(1986).

14 R.L. Jones, "Some Aspects of the Hot Corrosion of Thermal Barrier Coatings," J. of Thermal Spray Technology, 6, 1, 77-84 (1997).

15 M. Yoshiba, K. Abe, T. Arami, Y. Harada, High-Temperature Oxidation and Hot Corrosion Behavior of Two Kinds of Thermal Barrier Coating Systems for Advanced Gas Turbines, J. of Thermal Spray Technology, 5, 3, 259-68 (1996).

16 B. R. Marple, J. Voyer, M. Thibodeau, D. R. Nagy, R. Vaßen, Hot Corrosion of Lanthanum Zirconate and Partially Stabilized Zirconia Thermal Barrier Coatings, Transactions of the ASME. Journal of Engineering for Gas Turbines and Power, , 128, 1, 144-52 (Jan. 2006), ASME, Journal Paper. (AN: 8664377).

17 P.E. Hodge, R.A. Miller, M.A. Gedwill, Evaluation of the Hot Corrosion Behaviour of Thermal Barrier Coatings Thin Solid Films, 73, 447-453 (1980).

18 I. Zaplatynsky, Performance of Laser-Glazed Zirconia Thermal Barrier Coatings in Clyclic Oxidation and Corrosion Burner Rig Tests, Thin Solid Films, 95, 275-284 (1982).

19 P. Kofstad, High Temperature Corrosion, Elsevier Appl. Sci., London, 1988.

20 R. Vaßen, D. Sebold, G. Pracht, D. Stöver, Corrosion rig testing of thermal barrier coating systems, on the 30th Int. Cocoa Beach Conf. & Exposition, 23th -27th January 2006, Cocoa Beach, Fl..

21 F. Traeger, R. Vaßen, K.-H. Rauwald, D. Stöver, A Thermal Cycling Setup for Thermal Barrier Coatings, Adv. Eng. Mats., 5, 6, 429-32 (2003).

Microstructural
Characterization of
Thermal Barrier Coatings

MONITORING THE PHASE EVOLUTION OF YTTRIA STABILIZED ZIRCONIA IN THERMAL BARRIER COATINGS USING THE RIETVELD METHOD

G. Witz, V. Shklover, W. Steurer
Laboratory of Crystallography, ETH Zürich,
Wolfgang-Pauli-Str. 10
Zürich, 8093 Zürich, Switzerland

S. Bachegowda, H.-P. Bossmann
Alstom (Schweiz) AG
Brown Bovery Str. 7
Baden, 5401 Baden, Switzerland

ABSTRACT

Plasma sprayed thermal barrier coating composed of tetragonal Yttria Stabilized Zirconia (YSZ) have a limited lifetime and after a certain operating time they fail, usually by spallation. One of the proposed failure mechanisms is the transformation of metastable tetragonal YSZ phase into its monoclinic polymorph, which can lead to destabilization of coating structure and increase of thermal conductivity. Study of the monoclinic content in YSZ is usually performed by X-Ray diffraction using evaluation of the intensities of a few diffraction peaks for each YSZ phase. But this method is missing some important information that can be further extracted from the X-Ray diffraction pattern using Rietveld method. Using the Rietveld method, one can estimate phase content more accurately, take into account some phases present at level even below 1%, gain information on the grain size and strains. By applying Rietveld method, we can observe that during the first stage of the coating ageing process (a) small grains of cubic YSZ are crystallizing and (b) yttria content within the tetragonal phase is lowering. The tetragonal phase can be described as a mixture of two tetragonal phases; one (t') with typical for YSZ c/a ratio and unchanged yttria content and second one (t) with an increased c/a ratio, corresponding to a lower yttria content. These observations are used for modeling the decomposition of the tetragonal YSZ, considering the segregation of yttria at YSZ phase boundaries, leading to the formation of YSZ domains with the cubic structure.

INTRODUCTION

Thermal barrier coatings (TBC) are used in gas turbines to reduce the temperature of blades and hot engine components, they allow operating them at higher temperature, increase the components lifetime and reduce the cooling needs. A TBC system typically consists of a MCrAlY bond coat and an yttria stabilized zirconia (YSZ) topcoat. The topcoat is in general composed of metastable tetragonal YSZ containing 6-8 wt% of yttria. As can be seen in Figure 1, by varying the yttria content, other polymorphs of YSZ can be obtained.[1] Upon annealing at temperature above 1000 °C, the metastable tetragonal YSZ phase is expected to decompose into a mixture of cubic YSZ having a high yttria content and tetragonal YSZ having a low yttria content. The low yttria content tetragonal phase is stable only at high temperature and it undergoes a martensitic transformation to a monoclinic phase at a temperature between 600 °C and 1000 °C depending on the yttria content accompanied by a change of the unite cell volume by about 4%, which can result in cracking and coating failure.

Further improvement of engine efficiency and reduction of operating costs are expected by increasing the operating temperature and maintenance intervals.[2] Both of these goals require a better understanding of the mechanisms leading to YSZ coating failure. Decomposition of the metastable tetragonal polymorph of YSZ into cubic and monoclinic polymorphs has been considered as a potential mechanism for coating failure.[3] A spinodal decomposition model proposed by Katamura et al. dicates that yttrium diffuses to the domain boundaries, which have a cubic structure.[4] This leads to the formation of yttria depleted domains., which at some point of the decomposition process are sufficiently depleted in yttria to undergo the tetragonal-monoclinic transition upon cooling leading to volume change and potentially to coating failure.

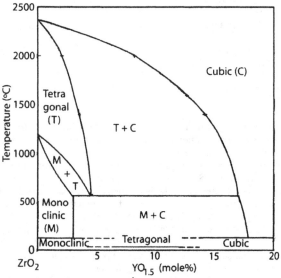

Fig. 1. Phase diagram of YSZ[1]

X-ray diffraction (XRD) combined with Rietveld refinement is an efficient technique to determine microstructural parameters and phase content of multiphase compounds. The Rietveld method uses a least square approach to fit a measured XRD pattern with a theoretical one. Microstructural and instrumental parameters, like cell dimensions, atomic positions and occupancies, phase composition, peak shapes including microstrains and grain size contributions, texture, polarizations effects, 2θ offset, sample displacement, absorption factors, atomic displacement factors and background contribution can be refined. Nevertheless, the Rietveld refinement of multiphase systems can be difficult if there are overlapping peaks like in YSZ polymorphs, in spite of using a monochromatic radiation that generally increases the quality of the collected spectra and reduces the peak overlaps. The X-rays scattering factors of Zr and Y are too close to allow refining directly their relative content in the unit cell. But since the lattice parameters of the tetragonal and cubic phases depend on the yttria content,[1,5] it is possible to determine the yttria content in both phases and study its evolution.

A previous neutron scattering study by Ilavsky et al. of the phase evolution in YSZ coatings during annealing shows that metastable tetragonal YSZ decomposes into monoclinic and cubic phases while the yttria content of the tetragonal phase is reduced.[5] XRD studies on plasma sprayed coatings by Brandon and Taylor and on EB-PVD coatings by Schulz indicate that at 1300 °C and 1400 °C the metastable tetragonal phase decomposes into a new low-yttria content tetragonal phase and a high-yttria content cubic or tetragonal phase.[6,7] Computer simulation, combined with TEM studies has shown that the domain boundaries have a cubic-like structure and that yttrium ions tend to concentrate in these boundaries.[4]

In the present work, the microstructure evolution in plasma sprayed YSZ-TBC during annealing at temperature ranging from 1100 °C to 1400 °C is studied. Lattice parameters, phases content, strains and crystallite size of as sprayed and annealed coatings are determined using Rietveld refinement of XRD patterns.

EXPERIMETAL PROCEDURE

YSZ coatings were produced by an atmospheric plasma spraying system (APS). The feedstock material is a zirconia with 7.8 wt% yttria powder (204NS, Sulzer-Metco, Westbury, NY). Samples were mechanically removed from the substrate before annealing at temperatures ranging from 1100 °C to 1400 °C. After annealing samples were furnace cooled down to room temperature such that the transformation from tetragonal to monoclinic YSZ can occur during cooling. XRD patterns were collected using an X-ray powder diffractometer (X'Pert Pro, PANalytical, Almelo, The Netherlands) with monochromatic CuKα_1 radiation, within 2θ=20 to 100° range in 0.01° steps using standard θ–2θ Bragg–Brentano geometry. Rietveld refinement was performed using the General Structure Analysis System (GSAS) package and the EXPGUI interface.[8,9] The Rietveld refinement was processed following the Rietveld refinement guidelines formulated by the International Union of Crystallography Commission on Powder Diffraction.[10] The background was fitted using Chebyschev polynomials and the peak profiles were fitted using a convolution of a pseudo-Voigt and asymmetry function together with the microstrain broadening description of P. Stephens.[11-13] The pseudo-Voigt is defined as a linear combination of a Lorentzian and a Gaussian. The Gaussian variance of the peak, σ^2, varies with 2θ as:

$$\sigma^2 = U \tan^2 \theta + V \tan \theta + W + P/\cos^2 \theta \tag{1}$$

where U, V and W are the coefficients described by Cagliotti et al. and P is the Scherrer coefficient for Gaussian broadening.[14] The Lorentzian coefficient, γ, varies as:

$$\gamma = \frac{X}{\cos \theta} + Y \tan \theta \tag{2}$$

The P and X coefficients are related to the Scherrer broadening and gives information about crystallite dimensions. The U and Y coefficients are related to the strain broadening and allow estimating micro-strains within crystallites:

$$\varepsilon_G (\%) = \frac{\pi}{180} \sqrt{8 \ln 2 (U - U_i)} \tag{3}$$

$$\varepsilon_L(\%) = \frac{\pi}{180}(Y - Y_i) \qquad\qquad (4)$$

The surface roughness and porosity of the coating leads to absorption of the X-Ray beam and reduction of the diffracted intensities at low angles. If this effect is not taken into account during Rietveld refinement, the atomic displacement parameters are reduced and can even become negative having then no more physical meaning. To take into account the absorption effect, we used the function described by Pitschke et al.[15] Since atomic displacement parameters and absorption factors modify the profile intensities with a similar 2θ dependency and refining all of them together usually leads to unrealistic results, we fixed the atomic displacement parameters to standard values and refined only the absorption factors. The number of refined parameters during the last step of the refinement process and R_p factors indicating the quality of the fit compared to the experimental XRD pattern are listed in Table I, the R_p values are calculated after removal of the background contribution.

Table I. Data on annealing conditions and Rietveld refinement

Temperature °C	Annealing time h	Number of refined parameters	R_p (without background)
-	-	27	7.8%
1400	1	33	6.9%
1400	10	33	6.5%
1300	1	27	7.4%
1300	10	33	6.4%
1300	100	33	5.7%
1300	1000	32	7.3%
1200	1	27	7.6%
1200	10	27	8.3%
1200	1000	33	7.6%
1100	1	27	6.7%
1100	24	27	6.8%
1100	100	27	6.0%
1100	250	33	6.3%
1100	450	33	6.1%
1100	650	33	6.1%
1100	850	33	6.6%
1100	1400	33	7.7%

Four YSZ phases were used for interpretation of the XRD patterns: two tetragonal phases with different unit cell dimensions corresponding to different yttria content of 7-8 wt% (t´-YSZ) and 4-5 wt% (t-YSZ), one monoclinic (m-YSZ) and one cubic phase having 13-15 wt% yttria (c-YSZ). For the coatings annealed at 1400 °C, the pattern of the phase with high yttria content (11-13 wt% yttria) was fitted using a tetragonal phase with a low c/a ratio, we will refer to this phase as c-YSZ to be consistent with the phase definitions used for lower annealing temperature. In some patterns, the intensities of peaks of some phases are too low to allow a structural refinement. In such cases, the unit cell dimensions are fixed by setting them to a defined yttria content. The peak shapes parameters are then either fixed or the number of refined parameters is reduced to 1 or 2. The yttria content of the cubic phase is calculated using the formula:

$$YO_{1.5}(mol\%) = (a - 5.1159)/0.001547 \qquad (5)$$

derived by Ilavsky[5] from the data of Scott[1], where a is the unit cell dimension in angstroms. The yttria content of the tetragonal phase is calculated using the formula:

$$YO_{1.5}(mol\%) = \frac{1.0225 - \frac{c}{a\sqrt{2}}}{0.0016} \qquad (6)$$

where a and c are the unit cell dimensions expressed in angstroms. This formula was derived from data of Scott[1] and empirically corrected by Ilavsky[16] to obtain a better fit with data coming from various samples.

RESULTS AND DISCUSSION

XRD patterns of the YSZ coating annealed at 1300 °C for various times are displayed in Figure 2. The pattern of the as-deposited coating contains only peaks belonging to t′-YSZ. After 10 hours of annealing, shoulders appear in Figure 2b at lower angles for the (004) peak at 2θ=73.2° and to higher angles for the (220) peak at 2θ=74.2°, they belong to the t-YSZ which has a higher c/a ratio. A new peak also appears between the (004) and (220) peaks of t′-YSZ indicating the presence of c-YSZ. After 100 hours of annealing, peaks of t-YSZ and c-YSZ grow when those of t′-YSZ are getting smaller, and after 1000 hours of annealing the peaks of t′-YSZ completely disappeared. In Figure 2a, peaks attributed to m-YSZ phase also appear after 1000 hours of annealing. The peak present at 2θ=30.2° is a combination of the (101) peaks of both tetragonal YSZ phases and the (111) peak of c-YSZ. This peak appears at the lowest angle for c-YSZ and at the highest angle for t′-YSZ, explaining its asymmetric shape.

These observations show that fitting XRD patterns using a combination of two tetragonal YSZ phases having different yttria content and unit cell dimensions is a good way to obtain a realistic description of the YSZ sample composition. Using only one unique tetragonal YSZ phase as was done by Ilavsky et al.[5] is not a good solution, since as can be observed in Figure 2b, peaks of the tetragonal phase are not shifted during annealing due to a steady loss of yttria, but a new set of peaks appears while the peaks of the as-deposited YSZ disappear. Using more than two tetragonal YSZ phases did not improve the quality of the fit, the phase contents obtained for phases having intermediate yttria contents were always close to 0 wt%. This indicates that we do not observe YSZ phases having intermediate yttria content, but only the as-deposited phase or the decomposition products, as was already observed by Brandon and Taylor in plasma sprayed coatings annealed at 1300 °C and higher temperatures.[6]

Evolution of the phase composition of the coatings during annealing at different temperatures is displayed in Figure 3 and the phase composition, cell parameters and yttria content of each phase are listed in Table II and III. We always observe decomposition of t′-YSZ into c-YSZ and t-YSZ with a decomposition rate increasing by around one order of magnitude when the temperature is increased by 130 °C. After 1000 hours of annealing at 1300 °C the m-YSZ appears; at the same time the peaks of t′-YSZ are no more visible and the amount of the t-YSZ starts to decrease.

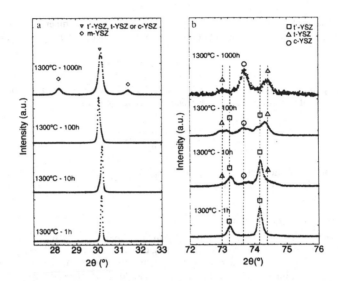

Fig. 2. a) XRD patterns of YSZ coatings, annealed for various times at 1300 °C, for a 2θ range between 27° and 33°. b) XRD patterns of YSZ coatings, annealed for various times at 1300 °C, for a 2θ range between 72° and 76°.

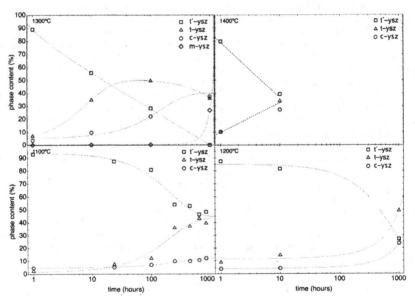

Fig. 3. Evolution of phase composition of YSZ coatings during annealing at different temperatures

Table II. Data on YSZ phases obtained from Rietveld refinement

Temperature (°C}	Annealing time (h)	t'-YSZ wt %	t-YSZ wt %	c-YSZ wt %	m-YSZ wt %	a t'-YSZ (Å)	c t'-YSZ (Å)	a t-YSZ (Å)	c t-YSZ (Å)	c c-YSZ (Å)
-	-	88.8	7.5	3.7	-	3.6150	5.1617	3.6044	5.1709	5.1325
1400	1	79.6	10.0	10.4	-	3.6132	5.1654	3.6006	5.1780	5.1467/ 3.6246
1400	10	39.3	33.5	27.2	-	3.6109	5.1634	3.6005	5.1783	5.1456/ 3.6248
1300	1	89.2	7.1	3.7	-	3.6144	5.1664	3.6045	5.1709	5.1353
1300	10	52.3	38.4	9.3	-	3.6134	5.1647	3.6057	5.1687	5.1353
1300	100	28.3	49.6	22.1	-	3.6138	5.1665	3.6029	5.1741	5.1361
1300	1000	-	35.9	37.4	26.7	-	-	3.6066	5.1774	5.1395
1200	1	87.2	8.9	3.8	-	3.6142	5.1669	3.6045	5.1709	5.1353
1200	10	81.5	14.3	4.2	-	3.6142	5.1671	3.6045	5.1709	5.1353
1200	1000	27.1	49.2	23.7	-	3.6143	5.1685	3.6031	5.1823	5.1403
1100	1	93.1	2.3	4.6	-	3.6121	5.1638	3.6041	5.1689	5.1373
1100	24	87.4	7.4	5.2	-	3.6125	5.1643	3.6041	5.1689	5.1373
1100	100	80.9	12.1	7.0	-	3.6118	5.1647	3.6041	5.1689	5.1373
1100	250	54.2	35.9	9.9	-	3.6116	5.1645	3.6041	5.1689	5.1373
1100	450	53.0	37.1	9.9	-	3.6117	5.1650	3.6041	5.1689	5.1373
1100	650	46.4	43.1	10.6	-	3.6108	5.1632	3.6041	5.1689	5.1373
1100	850	48.4	39.5	12.1	-	3.6117	5.1657	3.6041	5.1689	5.1373
1100	1400	42.3	45.0	12.7	-	3.6102	5.1638	3.6032	5.1672	5.1373

Curves describing the time evolution of t'-YSZ decomposition level can be scaled such to fall on a master curve. To do this, time scale was modified with an equation of the form:

$$t^* = At \cdot e^{-\frac{B}{T}} \qquad (7)$$

where t is the annealing time and T is the temperature in Kelvin. The master curve is displayed in Figure 4. The time scaling used allow that all data fall well on the same curve and it allows to estimate how fast t'-YSZ will decompose at a given temperature.

Information about micro-strains and grain size was extracted from the function describing the Gaussian part of the diffraction peak shape. It indicates that peak broadening is mainly governed by micro-strains within the crystallites. The micro-strain level in t'-YSZ is around 0.2-0.4% for short annealing times and increases up to 1% upon further annealing as can be seen in Figure 5. After 1 hour of annealing the micro-strains in the coatings annealed at 1300 °C and 1400 °C are lower than in the coating annealed at lower temperatures, this can be explained by the high level of micro-strains of 0.48% in the as-deposited coating that are released during the first hours of annealing at high temperature. The general subsequent increase in micro-strains can be related to the reduction in t'-YSZ phase content leading to an increase in straining due to the other phases present in the coating. In the c-YSZ and t-YSZ phases, micro-strains vary between 0.4% and 0.7% and decrease upon annealing. The micro-strain level that is reached in the t'-YSZ phase upon annealing at high temperature is high enough to allow cracks nucleation and their subsequent propagation. But it is probably overestimated because the micro-strain broadening is described by a distribution of cell dimensions; in the case of YSZ, this distribution can also come from a non-homogeneous yttria distribution within the grains or between the grains. Therefore, the increase of the observed micro-strains could also be interpreted as an effect of yttria diffusion, which leads to a broader distribution of the yttria content within t'-YSZ crystallites.

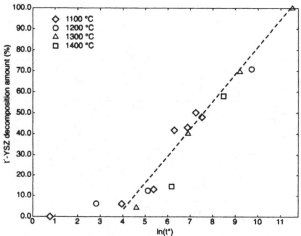

Fig. 4. Master curve of t'-YSZ content as a a function of equivalent annealing time at 1100 °C.

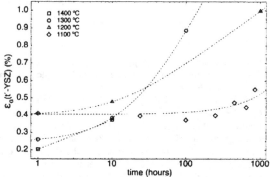

Fig. 5. Evolution of micro-strains in the t'-YSZ phase as a function of annealing time.

The absence of t′-YSZ undergoing the decomposition process indicates that the diffusion of yttrium should happen on a short time scale. This observation together with the diffraction peak profiles showing that the broadening comes mainly from a high level of strains or a non-homogeneous yttria distribution in the particles, are in agreement with previous TEM studies by Shibata et al.[17] and the spinodal decomposition model proposed by Katamura et al.[4] According to this model, yttrium segregates at domain boundaries producing grains composed of domains having the cubic structure with high yttria content and other domains having the tetragonal structure and a reduced yttria content. Only when the size of yttria-depleted domains reaches some critical size, they undergo the martensitic transformation to the monoclinic YSZ structure.[18] One can write the decomposition process as:

$$t′\text{-YSZ} \rightarrow t\text{-YSZ} + c\text{-YSZ} \rightarrow m\text{-YSZ} + c\text{-YSZ} \qquad (8)$$

It is still not clear if the remaining traces of t′-YSZ come from grains having a slower decomposition rate or if they are equally distributed in grains undergoing the decomposition process. It is also not yet clear if the m-YSZ is forming only after the t′-YSZ is fully decomposed. The answer to this question can be important for the consideration of the stabilization mechanism; it can gives information if t-YSZ domains can be stabilized by the presence of adjacent t′-YSZ domains or if the transformation from tetragonal to monoclinic YSZ is only governed by the size of the t-YSZ domains.

Table III. Data on YSZ phases obtained from Rietveld refinement

Temperature (°C)	Annealing time (h)	Yttria content t′-YSZ	Yttria content t-YSZ	Yttria content c-YSZ	U t′-YSZ	ε t′-YSZ
-	-	8.0%	5.0%	10.7%	135	0.48%
1400	1	7.3%	3.4%	13.2%	24.6	0.20%
1400	10	7.1%	3.4%	12.9%	82.8	0.38%
1300	1	7.4%	5.1%	12.5%	40.1	0.26%
1300	10	7.4%	5.0%	12.5%	87.9	0.39%
1300	100	7.2%	4.2%	13.1%	465	0.89%
1300	1000	-	4.6%	15.3%	-	-
1200	1	7.3%	5.0%	12.5%	99.2	0.41%
1200	10	7.2%	5.0%	12.5%	136	0.48%
1200	1000	7.1%	3.4%	15.8%	590	1.00%
1100	1	7.3%	5.3%	13.8%	101	0.42%
1100	24	7.3%	5.3%	13.8%	92.8	0.41%
1100	100	7.1%	5.3%	13.8%	82.1	0.40%
1100	250	7.1%	5.3%	13.8%	94.3	0.37%
1100	450	7.1%	5.3%	13.8%	133	0.40%
1100	650	7.1%	5.3%	13.8%	118	0.48%
1100	850	7.0%	5.3%	13.8%	180	0.55%
1100	1400	6.9%	5.3%	13.8%	191	0.57%

CONCLUSIONS

X-Ray diffraction combined with Rietveld refinement of full patterns allowed monitoring of evolution of phase composition in YSZ plasma sprayed coatings. When the coating is annealed at temperature ranging between 1100 °C and 1400 °C, the as-deposited tetragonal YSZ phase containing 7-8 wt% of yttria decomposes into a low yttria content tetragonal YSZ phase with 4-5 wt% of yttria and a high yttria content YSZ phase. This high yttria content YSZ phase is either cubic and contains 13-15 wt% of yttria when the annealing is performed at 1300 °C and lower temperatures or is tetragonal and contains 11-13 wt% of yttria when the annealing is performed at 1400 °C. When the domins of the low yttria content tetragonal YSZ phase have grown enough in size they transform into monoclinic YSZ upon colling. The kinetic of the phase transformation depends on the annealing temperature and is increased by around one order of magnitude when the temperature is increased by 130 °C. The decomposition of the as-deposited metastable YSZ phase comes together with an increase of its micro-strains having values high enough to initiate cracks that after propagation would lead to the coating failure.

REFERENCES

[1]H. G. Scott, "Phase relationships in the zirconia-yttria system", *J. Mater. Sci.*, **10**, 1527-35 (1975)

[2]D. R. Clarke and C. G. Levi, "Material Design for the Next Generation Thermal Barrier Coatings", *Annu. Rev. Mater. Res.*, **33**, 383-417 (2003)

[3]R. A. Miller, J. L. Smialek, and R. G. Garlick, "Phase Stability in Plasma-Sprayed, Partially Stabilized Zirconia–Yttria"; pp. 241–53 in Advances in Ceramics, Vol. 3, *Science and Technology of Zirconia I*. Edited by A. H. Heuer and L. W. Hobbs. American Ceramic Society, Columbus, OH, 1981

[4]J. Katamura and T. Sakuma, "Computer Simulation of the Microstructural Evolution during the Diffusionless Cubic-to-Tetragonal Transition in the System ZrO2-Y2O3", *Acta mater.*, **46** [5], 1569-75 (1998)

[5]J. Ilavsky, J. K. Stalick, and J. Wallace, "Thermal Spray Yttria-stabilized Zirconia Phase Changes during Annealing", *J. Therm. Spray Technol.*, **10** [3], 497-501 (2001)

[6]J. R. Brandon and R. Taylor, "Phase stability of zirconia-based thermal barrier coatings Part I, Zirconia-yttria alloys", *Surf. Coat. Technol.*, **46**, 75-90 (1991)

[7]U. Schulz, "Phase Transformation in EB-PVD Yttria Partially Stabilized Zirconia Thermal Barrier Coatings during Annealing", *J. Am. Ceram. Soc.*, **83** [4], 904 –10 (2000)

[8]A. C. Larson and R. B. Von Dreele, "General Structure Analysis System (GSAS)", *Los Alamos National Laboratory Report* LAUR 86-748 (2000)

[9]B. H. Toby, "EXPGUI, a graphical user interface for GSAS", *J. Appl. Cryst.*, **34**, 210-13 (2001)

[10]L. B. McCusker, R. B. Von Dreele, D. E. Cox, D. Louër, and P. Scardi, "Rietveld refinement guidelines", *J. Appl. Cryst.* **32**, 36-50 (1999)

[11]P. Thompson, D. E. Cox and J. B. Hastings, J. Appl. Cryst., "Rietveld refinement of Debye-Scherrer synchrotron X-ray data from Al2O3", *J. Appl. Cryst.*, **20**, 79-83 (1987)

[12]L. W. Finger, D. E. Cox and A. P. Jephcoat, "A correction for powder diffraction peak asymmetry due to axial divergence", *J. Appl. Cryst.*, **27**, 892-900 (1994)

[13]P. Stephens, "Phenomenological model of anisotropic peak broadening in powder diffraction", *J. Appl. Cryst.*, **32**, 281-89 (1999).

[14]G. Caglioti, A. Paoletti and F. P. Ricci, "Choice of collimators for a crystal spectrometer for neutron diffraction", *Nucl. Instrum.*, **3**, 223-28 (1958)

[15]W. Pitschke, H. Hermann, and N. Mattern, "The influence of surface roughness on diffracted X-ray intensities in Bragg-Brentano geometry and its effect on the structure determination by means of Rietveld analysis", *Powder Diffr.*, **8**, 74-83 (1993)

[16]J. Ilavsky and J. K. Stalick, "Phase composition and its changes during annealing of plasma-sprayed YSZ", *Surf. Coat. Technol.*, **127** [2-3], 120-29 (2000)

[17]N. Shibata, J. Katamura, A. Kuwabara, Y. Ikuhara, T. Sakuma, "The instability and resulting phase transition of cubic zirconia", *Mater. Sci. Eng.*, **A312**, 90–8 (2001)

[18]T. K. Gupta, F. F. Lange, J. H. Bechtold, "Effect of stress-induced phase transformation on the properties of polycrystalline zirconia containing metastable tetragonal phase", *J. Mat. Sci.* **13**, 1464-70 (1978)

THERMAL IMAGING CHARACTERIZATION OF THERMAL BARRIER COATINGS

J. G. Sun
Argonne National Laboratory
Argonne, IL 60439

In a three-layer thermal barrier coating (TBC) system consisting of a ceramic TBC topcoat, a bond coat, and a metallic substrate, a large disparity in thermal conductivity exists between the TBC and the substrate and, when TBC is debonded and air fills the gap, between the TBC and the air. For TBC system characterization, flash thermal imaging is effective because it involves nondestructive measurement of thermal properties. This paper describes a new thermal-imaging method for multilayer TBC characterization and imaging to simultaneously determine the TBC thickness, conductivity, and optical absorptance. This method directly accounts for the TBC translucency that has been a major issue in thermal imaging application for TBCs. Results from theoretical analyses and experimental measurements are presented and discussed.

INTRODUCTION

Thermal barrier coatings (TBCs) are being extensively used for improving the performance and extending the life of combustor and gas turbine components. In this application, a thermally insulating ceramic topcoat (the TBC) is bonded to a thin oxidation-resistant metal coating (the bond coat) on a metal substrate. Because TBCs play critical role in protecting the substrate components, their failure (spallation) may lead to unplanned outage or safety threatening conditions. Therefore, it is important to inspect and monitor the TBC condition to assure its quality and reliability.

Most TBCs consist of yttria-stabilized zirconia (YSZ). They are usually applied by electron-beam physical vapor deposition (EB-PVD) or air plasma spay (APS) to thicknesses ranging from 0.1 to >2 mm. TBC failure during high-temperature operations typically starts from initiation of small cracks at the TBC/bond coat interface. These cracks then grow and link together to form delaminations under the TBC which will eventually cause the TBC spallation. To monitor this TBC failure process and detect TBC delamination, several nondestructive evaluation (NDE) methods have been developed, most are based on optical principles because TBCs are typically either semi-transparent (most EB-PVD TBCs) or translucent (APS TBCs). These optical NDE methods include mid-infrared reflectance [Eldridge et al., 2006], luminescence spectroscopy [Tolpygo et al., 2004], and elastic optical scattering [Ellingson et al., 2006]. These methods have been demonstrated to be capable of detecting TBC degradation and pre-spall conditions. However, they can only be used for semi-transparent EB-PVD TBCs or thin APS TBCs (<0.4 mm) because of the limited optical penetration depth of the lights used in these methods. In addition, these methods are qualitative and may become useless once a TBC is coated or infiltrated by "dirty" contaminants.

In the three-layer TBC system consisting of a TBC topcoat, a bond coat, and a metal substrate, a large disparity in thermal conductivity exists between the TBC and the substrate and, when TBC is delaminated and air fills the gap, between the TBC and the air. Therefore, pulsed (or flash) thermal imaging is effective for TBC system characterization because it involves nondestructive measurement of thermal properties. Thermal imaging has been widely used to detect TBC delamination [e.g., Chen et al., 2001]. Recently, it has also been extended for estimation of TBC thickness and thermal conductivity [Shepard et al., 2005; Ringermacher, 2004]. However, because TBC system is multilayer and its top layer (TBC) is translucent, conventional methods for pulsed thermal imaging cannot be used directly to analyze the TBC system. This paper describes a new thermal-imaging technology for multilayer TBC characterization and imaging. Results from experimental and theoretical analyses are presented and discussed.

PULSED THERMAL IMAGING METHODS FOR SINGLE- AND MULTI-LAYER MATERIALS

Pulsed thermal imaging is based on monitoring the temperature decay on a specimen surface after it is applied with a pulsed thermal energy that is gradually transferred inside the specimen. The premise is that the heat transfer from the surface (or surface temperature/time response) is affected by internal material structures and properties and the presence of flaws such as cracks [Sun, 2006a]. A schematic one-sided pulsed-thermal-imaging setup for testing a 3-layer material system is illustrated in Fig. 1. Theoretical development for analyzing material properties from thermal imaging data is described below.

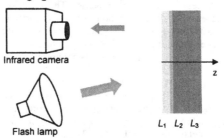

Fig. 1. Schematics of pulsed thermal imaging of a 3-layer material system.

Because thermal imaging is inherently a 2D method (it images the temperature of the 2D x-y specimen surface), theoretical development is usually carried out in 1D (the depth or z direction) models. The temperature/time response at a surface position (a pixel in a 2D image) is related to the depth variation of material properties under that surface position. By analyzing the surface temperature/time response, the material property and depth of various subsurface layers under that pixel can be determined. The final thermal imaging results for all pixels are usually presented in image forms; the value at each pixel represents a particular physical parameter such as thermal conductivity or depth that was determined from the analysis. To understand the thermal responses due to material thermal and optical properties and depth, heat conduction theory is examined first.

The 1D governing equation for heat conduction in a solid material is:

$$\rho c \frac{\partial T}{\partial t} = \frac{\partial}{\partial z}\left(k \frac{\partial T}{\partial z} \right), \tag{1}$$

where $T(z, t)$ is temperature, ρ is density, c is specific heat, k is thermal conductivity, t is time, z is coordinate in the depth direction, and $z = 0$ is the surface that receives pulsed heating. It is noted that Eq. (1) contains two independent thermal parameters, the heat capacity ρc and the thermal conductivity k, both are normally assumed constant in each material layer.

During flash thermal imaging, an impulse energy is applied on surface $z = 0$ at $t = 0$. Under ideal thermal imaging conditions which assumed (1) flash is instantaneous or flash duration is zero and (2) flash heat is absorbed on surface or heat-absorption depth is zero (for opaque materials), analytical solution of Eq. (1) for single-layer materials has been obtained by Parker et al. [1961]. Analytical solutions of Eq. (1) for single-layer materials under finite flash duration and finite heat-absorption depth (for translucent materials) were also obtained [Sun & Benz, 2004; Sun, 2006b]. These theories have been directly used for thermal imaging analysis of single-layer materials [Sun, 2006a, 2007].

Thermal imaging analysis for multilayer materials is more complex. For multilayer materials, parameters in each layer include: conductivity k, heat capacity ρc, layer thickness L, and, for translucent materials, the absorption coefficient a. In comparison, only one parameter α/L^2 ($\alpha = k/\rho c$) controls the entire heat transfer process in single-layer materials. Despite of the complexity, surface temperature decay for multilayer materials under pulsed thermography conditions has been well understood. For a 2-layer opaque material, depending on the ratio of heat conductivities between the first and second layers, k_1/k_2, the expected surface temperature decay is illustrated in Fig. 2 (in log-log scale). In the early time period, flash heat absorbed on the surface propagates within the first layer, and the surface temperature decay follows the -0.5 slope in log-log scale. When heat approaches the interface between the first and the second layer, the temperature decay rate deviates from the -0.5 slope if $k_1/k_2 \neq 1$. The temperature decay within this intermediate time period is therefore indicative of the interface condition. In later times, heat propagation proceeds in the second layer so the temperature decay rate is determined by the conductivity ratio k_1/k_2: when $k_1/k_2 < 1$ the (absolute) magnitude of the slope is >0.5, and when $k_1/k_2 > 1$ the (absolute) slope amplitude is <0.5. The surface temperature decay rate will eventually approaches to zero when the temperature of the entire specimen becomes equalized.

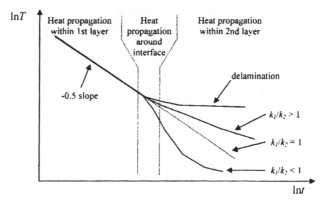

Fig. 2. Illustration of surface temperature decay from pulsed thermal imaging for a 2-layer material system.

For multilayer materials, direct solution of Eq. (1) is possible. Balageas et al. [1986] derived analytical solutions for two- and three-layer materials. However, these solutions are difficult to be used for general applications because a large number of eigenfunctions are involved. New robust and efficient methods are needed for thermal imaging analysis of multilayer material systems.

A general method and numerical algorithm has been developed for automated analysis of thermal imaging data for multilayer materials [Sun, 2006b]. It is based on a theoretical model of the material system which is solved numerically. The numerical formulation also incorporates finite flash duration and finite heat absorption depth effects. The numerical solutions are fitted with the experimental data by least-square minimization to determine unknown parameters in the multilayer material system. Multiple parameters in one or several layers can be determined simultaneously. For a TBC system, the most important parameters are the thickness, thermal conductivity, and absorption coefficient of the TBC in the first layer.

The least-square fitting is carried out for each pixel within the 2D thermal imaging data, and each fitted parameter is expressed in a 2D image. This data analysis process has been fully automated to simultaneously determine the distributions (images) of TBC thickness, conductivity, and absorption coefficient [Sun, 2006b]. Typical results for TBC analysis are presented below.

THERMAL IMAGING ANALYSIS OF TBC MATERIALS

For a TBC system, because the bond coat is typically thin and has thermal properties comparable to those in the substrate, thermal imaging analysis can be carried out for a two-layer material consisting of a TBC and a substrate. In this system, the conductivity of the TBC is typically much lower than that of the substrate, i.e., $k_1/k_2 < 1$. Therefore, the surface temperature decay should follow the curve for $k_1/k_2 < 1$ in Fig. 2 for flash thermal imaging of opaque TBCs. However, TBCs are typically translucent at levels determined by the amount of contamination. Because TBC translucency affects the heat absorption and infrared detection during a thermal imaging test, it must be determined explicitly in order for accurate prediction of other TBC parameters.

The numerical thermal imaging method described above is used to analyze a 2-layer TBC system to demonstrate its sensitivity and accuracy for determining TBC thickness, conductivity, and optical absorption coefficient. The results will be compared with thermal imaging data for a TBC specimen presented in the next section. In this 2-layer TBC system, the substrate is assumed to have constant thermal properties: $k_2 = 8$ W/m-K, $\rho c = 4$ J/cm^3-K, and thickness $L_2 = 2.5$ mm. The TBC has generic properties: $k_1 = 1.3$ W/m-K, $\rho c = 3$ J/cm^3-K, $L_1 = 0.62$ mm, and optical absorption coefficient $a = 4$ mm^{-1}.

Figure 3 shows the calculated results for TBCs with different optical absorption coefficient a. The TBC translucency (i.e., a finite a) can significantly reduce the temperature decay rate $d(\ln T)/d(\ln t)$ in the early times. When the absorption coefficient approaches infinity (for opaque TBCs), the initial surface-temperature slope becomes -0.5. Because TBC absorption depends on TBC material composition and structure as well as TBC surface conditions (such as contamination), this thermal imaging method can be used to analyze as-processed TBCs that have uniform optical property as well as used TBCs that may have various levels of surface contamination.

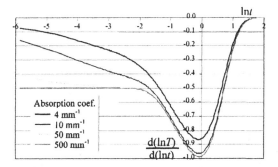

Fig. 3. Calculated surface-temperature slope as function of time (in log-log scale) for TBCs with various optical absorption coefficients.

Figure 4 shows calculated results for TBCs of different thicknesses. The surface-temperature slope $d(\ln T)/d(\ln t)$ initially follows approximately straight lines with (absolute) magnitudes below 0.5; the slope magnitude becomes larger than 0.5 when heat transfer reaches the substrate. The time when the slope change occurs is related to the TBC thickness.

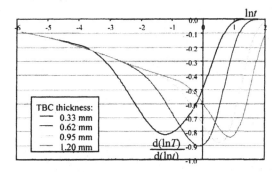

Fig. 4. Calculated surface-temperature slope as function of time (in log-log scale) for TBCs of different thicknesses.

Figure 5 shows calculated results for TBCs with different thermal conductivities. It is seen that the magnitude of the surface-temperature slope is very sensitive to the change of thermal conductivity of the TBC layer; a lower TBC conductivity will result in a higher peak magnitude of the slope.

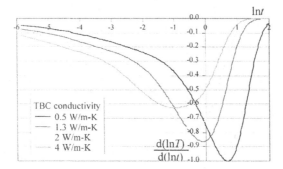

Fig. 5. Calculated surface-temperature slope as function of time (in log-log scale) for TBCs with different conductivities.

THERMAL IMAGING RESULTS FOR A TBC SPECIMEN

Pulsed thermal imaging test was conducted for an as-processed TBC specimen shown in Fig. 6a. It consists of a nickel-based substrate of 2.5 mm thick and a TBC layer with its surface being divided into 4 sections having nominal thicknesses 0.33, 0.62, 0.95, and 1.2 mm. Because this TBC specimen is as-processed, its thermal conductivity and optical absorption coefficient are expected to be uniform. Pulsed thermal imaging data was obtained for a total duration of 13 seconds at an imaging speed of 145 Hz. A typical thermal image is shown in Fig. 6b. Figure 7 shows measured surface-temperature slopes from the 4 thickness sections of this TBC specimen. Compared with the theoretical results in Fig. 4, the experimental data in Fig. 7 clearly indicate the difference of TBC thickness in these 4 sections.

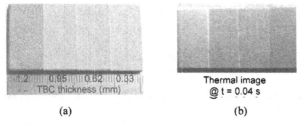

(a)   (b)

Fig. 6. (a) Photograph and (b) thermal image of a TBC specimen with 4 sections of thicknesses.

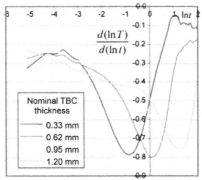

Fig. 7. Measured surface-temperature-slope data for TBCs of different thicknesses.

CONCLUSION
    A new thermal-imaging method for analyzing TBCs was developed. It is based on a theoretical model of the multilayer material system which is solved numerically. The numerical solution is fitted with experimental data by least-square minimization to determine unknown parameters. The method was used to analyze a 2-layer TBC system to demonstrate its sensitivity and accuracy for determining TBC thickness, conductivity, and optical absorption coefficient. The theoretical results agree well with thermal imaging data for a TBC specimen. This method has been fully automated to image TBC thickness, conductivity, and absorption coefficient distributions [Sun, 2006b]. Because it can accurately determine TBC thickness and conductivity variation, this method may be applied for health monitoring of TBC materials.

ACKNOWLEDGMENT
    Work sponsored by the U.S. Department of Energy, Energy Efficiency and Renewable Energy. Office of Industrial Technologies, Office of Power Technologies, under Contract W-31-109-ENG-38.

REFERENCES
    D. L. Balageas, J. C. Krapez, and P. Cielo, 1986, "Pulsed Photothermal Modeling of Layered Metarials." J. Appl. Phys.. Vol. 59, pp. 348-357.
    X. Chen, G. Newaz, and X. Han, 2001, "Damage Assessment in Thermal Barrier Coatings Using Thermal Wave Image Technique", Proc. 2001 ASME Int. Mech. Eng. Congress Expo., Nov. 11-16, 2001, New York, NY, paper no. IMECE2001/AD-25323.
    J. I. Eldridge, C. M. Spuckler, and R. E. Martin, 2006. "Monitoring Delamination Progression in Thermal Barrier Coatings by Mid-Infrared Reflectance Imaging," Int. J. Appl. Ceram. Technol., Vol. 3, pp. 94-104.
    W. A. Ellingson, R. J. Visher. R. S. Lipanovich, and C. M. Deemer, 2006. "Optical NDE Methods for Ceramic Thermal Barrier Coatings." Materials Evaluation. Vol. 64, pp. 45-51.

W. J. Parker, R. J. Jenkins, C. P. Butler, and G. L. Abbott, 1961, "Flash Method of Determining Thermal Diffusivity, Heat Capacity, and Thermal Conductivity," J. Appl. Phys., Vol. 32, pp. 1679-1684.

H. I. Ringermacher, 2004, "Coating Thickness and Thermal Conductivity Evaluation Using Flash IR Imaging," presented in Review of Progress in Quantitative NDE, Golden, CO, July 25-30, 2004.

S. M. Shepard, Y. L. Hou, J. R. Lhota, D. Wang, and T. Ahmed, 2005, "Thermographic Measurement of Thermal Barrier Coating Thickness," in Proc. SPIE, Vol. 5782, Thermosense XXVII, 2005, pp. 407-410.

J. G. Sun, 2006a, "Analysis of Pulsed Thermography Methods for Defect Depth Prediction," J. Heat Transfer, Vol. 128, pp. 329-338.

J. G. Sun, 2006b, "Method for Analyzing Multi-Layer Materials from One-Sided Pulsed Thermal Imaging," Argonne National Laboratory Invention ANL-IN-05-121, US patent pending.

J. G. Sun, 2007, "Evaluation of Ceramic Matrix Composites by Thermal Diffusivity Imaging," Int. J. Appl. Ceram. Technol., in press.

J. G. Sun and J. Benz, 2004, "Flash Duration Effect in One-Sided Thermal Imaging," in Review of Progress in Quantitative Nondestructive Evaluation, eds. D.O. Thompson and D.E. Chimenti, Vol. 24, pp. 650-654.

V. K. Tolpygo, D. R. Clarke, and K. S. Murphy, 2004, "Evaluation of Interface Degradation during Cyclic Oxidation of EB-PVD Thermal Barrier Coatings and Correlation with TGO Luminescence," Surf. Coat. Technol., Vol. 188-189, pp. 62-70.

EXAMINATION ON MICROSTRUCTURAL CHANGE OF A BOND COAT IN A THERMAL BARRIER COATING FOR TEMPERATURE ESTIMATION AND ALUMINUM-CONTENT PREDICTION

Mitstutoshi Okada and Tohru Hisamatsu
Central Research Institute of Electric Power Industry
2-6-1 Nagasaka
Yokosuka, 240-0196, Japan

Takayuki Kitamura
Kyoto University
Yoshida-honmachi
Kyoto, 606-8501, Japan

ABSTRACT

Specimens of superalloy with thermal barrier coating (TBC) are exposed to high-temperature atmosphere in order to develop a prediction method for local temperature and Al-content at bond coat (BC). The Al-content measured by means of an electron probe microanalyzer decreases as the test time passes. It is due to the Al transport induced by the oxidation of BC and the interdiffusion between BC and substrate. The Al-decreased layer (ADL) is formed at the boundary between BC and top coat since Al diffuses to the BC surface for the oxidation. Its thickness increases in proportion to the square root of test time, and the growth rate follows the Arrhenius relationship. Based on this relation, the local temperature of an in-service blade can be estimated by the ADL thickness if the operation period is known. The decrease of Al-content is also in proportion to the square root of test time, and Arrhenius relationship is established for the decrease rate. The prediction method of the Al-content is presented.

INTRODUCTION

In order to increase thermal efficiency of a gas turbine for electric generation, its turbine inlet temperature (TIT) has reached 1500°C at present [1]. Particularly, hot-gas-path parts such as combustors, vanes and blades are exposed to combustion gas flow, which is critical environment. For their reliability and reduction of maintenance cost, development of life evaluation method is inevitable.

Thermal barrier coating (TBC) as well as the internal cooling plays an important role as the gas temperature increases. Since it is difficult to measure the surface temperature of the hot-gas-path parts, the accurate estimation of the temperature distribution is important for life evaluation. Several estimation methods based on the microstructural change of substrates or coatings have been proposed. The temperature estimation method based on the diameter of γ' precipitate was presented for Ni-base superalloys, which are widely used for turbine blades [2-4]. However its growth rate changes due to the coalescence with neighboring γ', the influence on the temperature estimation has not been examined. On the other hand, an estimation method based on the microstructural change of coating (corrosion-resistant coating, MCrAlY) was proposed [2, 5]. Although the material for BC is often almost same as that for the corrosion-resistant coating, the estimation method by means of its microstructural change has not been developed. For the TBC, the methods focused on the oxide of BC (TGO) [6, 7] and top coat porosity [8] have been

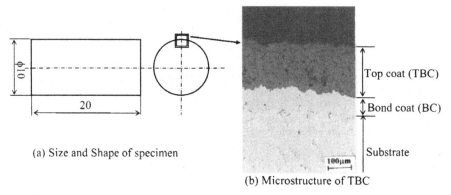

(a) Size and Shape of specimen

Top coat (TBC)

Bond coat (BC)

Substrate

(b) Microstructure of TBC

Figure 1. Schematic representation of specimen.

Table I. Chemical composition of Inconel738LC

wt%

C	Si	Mn	P	S	Ni	Cr	Mo	Co
0.09	0.02	0.01	<0.005	0.001	Bal.	16.00	1.70	8.48

W	Al	Ti	Fe	Ta	Cu	Ag	Bi
2.54	3.52	3.45	0.06	1.74	<0.01	<0.5 ppm	<0.1 ppm

reported. Both of them, however, have not been examined in terms of applicable limits and accuracy.

The BC oxidation causes TBC delamination under high-temperature environment [9-12]. Particularly, the decrease of Al-content in BC accelerates the delamination since it promotes the oxidation. There are few researches on the microstructural change due to Al diffusion caused by the oxidation [13-15].

This paper clarifies the microstructural change of BC, and proposes an estimation method of local temperature. The decrease of Al-content in BC, moreover, is examined.

EXPERIMENTAL PROCEDURE
Specimen
Figure 1(a) shows the size and shape of specimen. The cylindrical substrate, 10mm in diameter and 20mm in length, is made of an Inconel738LC, which is a typical material for gas turbine blade. Table I indicates its chemical composition. Figure 1(b) represents the microstructure of TBC. BC of CoNiCrAlY (Co-32Ni-21Cr- 8Al-0.5Y (wt %)) with the thickness of 100μm is formed on the substrate by the low pressurized plasma spraying (LPPS). The heat treatment is carried out after the spraying at 1393 K × 2h and 1118K × 24 h in a vacuum. Then, top coat (TBC) of yttria partially-stabilized zirconia (YSZ, 8wt% Y_2O_3-ZrO_2) with the thickness of 200μm is deposited by the air plasma spraying (APS).

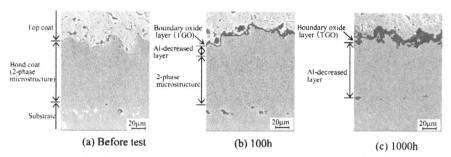

(a) Before test   (b) 100h   (c) 1000h

Figure 2. Microstructural change of bond coat in TBC specimens at 1273K in air.

Experimental procedure

The test is carried out at constant temperatures of 1173K, 1223K, 1273K and 1323K in an air by means of an electric tube furnace with the internal diameter of 70mm. The specimen is heated at 200K/h before the test, and the temperature fluctuation in the test section is kept in about 2~3K during the test. After the test, the specimen is cooled to about 773K at the rate of 200K/h and to the room temperature inside the furnace without the temperature control. Then, the microstructure of specimen is examined by means of an optical microscope, a scanning electron microscope (SEM) and an electron probe microanalyzer (EPMA).

TEMPERATURE ESTIMATION METHOD BY MICROSTURUCTURAL CHANGE OF BOND COAT

Figure 2 shows the microstructural change of BC at 1273K. The boundary between TBC and BC has the asperity of about 10μm, and the oxide (boundary oxide layer) grows along it. The oxide thickness increases as the test time passes.

The BC originally consists of 2-phases; dark dots in bright mother-phase as shown in Figure 2(a). The microstructure disappears from the BC surface near the TBC/BC boundary as shown in Figures 2(b) and (c). On the other hand, no significant microstructural change is observed at BC/substrate boundary. Figure 3 indicates the distribution of elements around the TBC/BC boundary before the test and after 1223Kx500h observed by an EPMA. It clearly points out that the 2 phases are β-(Ni, Al) and γ-(Co, Cr), and aluminum oxide is formed at the boundary. The lower Al-content region "Al-decreased layer (ADL)", which is caused by the diffusion, is observed on the boundary between the oxide layer and 2-phase microstructure in the BC. The Al-content in ADL is about 4wt%.

Figure 4 shows the relationship between the test time and the ADL thickness, which is the average of 12 data points in each specimen. The average squared root of the unbiased variance of the thickness is smaller than 10μm in all the test conditions. The ADL thickness increases monotonously according to the test time and temperature. The ADL grows to the whole bond coat after 500h at 1273K and 200h at 1323K. The relationship between the ADL thickness, l (μm) and the test time t (h) is in the form:

$$l = kt^{\frac{1}{2}}$$

(1)

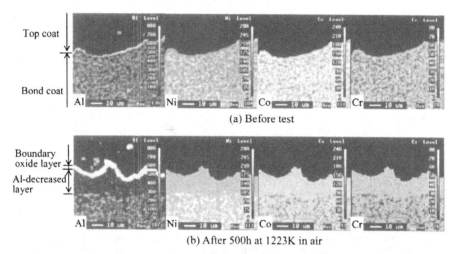

(a) Before test

(b) After 500h at 1223K in air

Figure 3. Distribution of elements around the boundary between top coat and bond coat by means of EPMA.

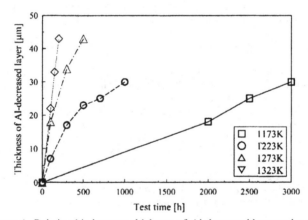

Figure 4. Relationship between thickness of Al-decreased layer and test time.

where k is the constant representing the growth rate. The Arrhenius plot shown in Figure 5 indicates

$$k = 2.96 \times 10^6 \exp\left(\frac{-152 \times 10^3}{RT}\right)$$

(2)

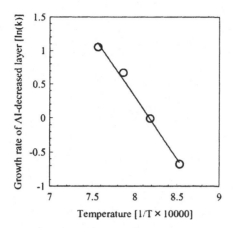

Figure 5.   Arrhenius plot growth rate of Al-decreased layer.

where T and R are temperature (K) and gas constant (8.31J/(mol·K)), respectively. The activation energy for the layer growth is given by 152kJ/mol. Then, T is described as follows from the equations (1) and (2).

$$T = \varphi(l,t) = -\frac{152 \times 10^3}{R} \cdot \frac{1}{\ln \dfrac{l}{2.96 \times 10^6 t^{1/2}}} \tag{3}$$

Equation 3 enables us to estimate the temperature at the vicinity of the BC surface by the measurement of ADL thickness at the hot section of in-service component if the operation time is known.

The detection limit of ADL by an optical microscope is about 10μm. It is the minimum thickness to be recognized as layer since the thickness has scatter. When the thickness is larger than about 50μm, the thickness cannot be distinguished. The ADL larger than 50μm coalescences with another ADL growing at vicinity of the boundary between the BC and the substrate due to the interdiffusion. Thus, the ADL thicknesses from 10μm to 50μm are the applicable condition of the method. As the applicable limits depend on the temperature, the applicable operation time varies as shown in Figure 6. The scatter of data caused by the asperity at the BC surface as shown in Figure 2 affects the accuracy as well.

Assuming that the error of operation time is neglected, the unbiased variance of temperature u_t^2 is expressed as follows [16].

$$u_t^2 = \left(\frac{\partial \varphi}{\partial l}\right)_0^2 \frac{u_l^2}{n} \tag{4}$$

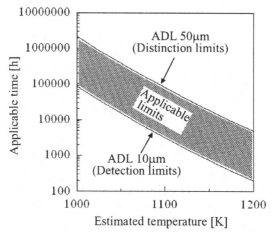

Figure 6. Applicable limits of the temperature estimation method.

Here, $\left(\frac{\partial \varphi}{\partial l}\right)_0$ is the differential of equation (3) by the ADL thickness l. u_l and n are the the square root of unbiased variance of ADL thickness and the number of the measurement, respectively. Let $u_l = 10$, $n = 10$, u_T is calculated. u_T varies with temperature and time since the ADL thickness is their function. Figure 7 shows the relationship between the estimated temperature and its error $\lambda \frac{u_l}{\sqrt{n}}$ (confidence coefficient 99%). The estimation error is smaller than 20K in almost all applicable limits.

Al-CONTENT PREDICTION METHOD

Figure 8 shows the relationship between the decrease of Al-content in BC, Δc, and square root of oxidation time. The Al-content is average of the whole bond coat. When it reaches about 4wt%, the diffusion saturates. Thus, these data are removed from Figure 8. The relationship is in the form;

$$\Delta c = c_0 - c = k' \sqrt{t} \tag{5}$$

The Arrhenius plot shown in Figure 9 indicates

$$k' = 8.93 \times 10^4 \exp\left(-\frac{142 \times 10^3}{RT}\right) \tag{6}$$

Thus, the time t is formulated as,

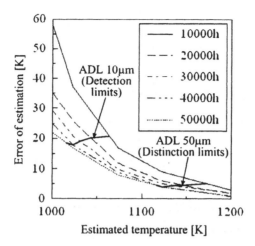

Figure 7. Error of estimated temperature by means of Al-decreased layer
(Confidence coordinate 99%).

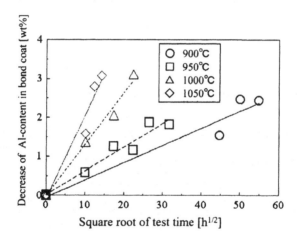

Figure 8. Relationship between the decrease of Al-content in bond coat and square root of test
time

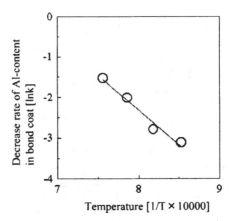

Figure 9. Arrhenius plot of Al-content decrease rate.

$$t = \left(\frac{c_0 - c}{8.93 \times 10^4 \exp\left(-\dfrac{142 \times 10^3}{RT}\right)} \right)^2 \tag{7}$$

Substituting the estimated temperature T evaluated by the ADL thickness into the equation (7), the time when the Al-content reaches an arbitrary value c is predicted. This method, however, can be used for the Al-content from about 8wt% to 4wt %.

CONCLUSION
    The Al-decreased layer is formed at the vicinity of the bond coat surface in the TBC. Its thickness increases in proportion to square root of the test time. The operation temperature can be estimated by measuring the thickness of the Al-decreased layer in a blade of in-service gas turbine on the basis of the relation among the thickness, time and temperature. The applicable condition due to the thickness of BC and detection limit of ADL is given in Figure 6.
    The Al-content of the bond coat is measured by means of an electron probe microanalyzer. The decrease of the Al-content is also in proportion to square root of the time. Using the relationship, the time to reach arbitrary Al-content can be predicted.

REFERENCES
[1]T. Okubo, "1500°C Class Steam Cooled Gas Turbine Combined Cycle Technology", Journal of the Gas Turbine Society of Japan", 31, 161-166 (2003)
    [2]V. Srinivasan, N. S. Cheruvu, T. J. Carr and C. M. O'Brien, "Degradation of MCrAlY Coating and Substrate Superalloy During Long Term Thermal Exposure", Materials and Manufacturing Process. 10, 955-969 (1995)

[3] Y. Yomei, N. Okabe, D. Saito, K. Fujiyama and T. Okamura, "Service Temperature Estimation of Gas Turbine Buckets Based on Microstructural Change", Journal of the Society of Materials Science , Japan, **45**, 699-704 (1996)

[4] A. Nomoto, M. Yaguchi and T. Ogata, "Evaluation of Creep Properties of Directionally Solidified Nickel Base Superalloy for Gas Turbine Blades Based on Microstructures", Central Research Institute of Electric Power Industry Report, T99094 (2000)

[5] M. Okada, Y. Etori, T. Hisamatsu and T. Takahashi, "Temperature estimation and prediction of Aluminum –content by means of microstructural change in gas turbine coatings". Journal of the Society of Materials Science , Japan, **54**, 257-264 (2005)

[6] M. Arai and U. Iwata, "Temperature estimation of gas turbine combustor based on thermally grown oxidation measurement in thermal barrier coating", The Thermal and Nuclear Power, **54**, 1064-1069 (2003)

[7] T. Torigoe, S. Aoki, I. Okada and H. Koguma, "Metal temperature estimation of high temperature components", JP.2003-4548 (2003).

[8] T. Fujii and T. Takahashi, "Development of Operating Temperature Prediction Method Using Thermophysical Properties Change of Thermal Barrier Coatings", Journal of Engineering for Gas Turbine and Power, **126**, 102-106 (2004)

[9] R. A. Miller, "Oxidation-Based Model for Thermal Barrier Coating Life", Journal of the American Ceramic Society, **67**, 517-521 (1984)

[10] S. Bose and J. DeMasi-Marcin, "Thermal Barrier Coating Experience in Gas Turbine Engines at Pratt & Whitney", Journal of Thermal Spray Technology, **6**, 99-104 (1997)

[11] A. Rabiei and G. Evans, Failure Mechanism Associated with the Thermally Grown Oxide in Plasuma-sprayed Thermal Barrier Coatings, Acta materialia., **48**, 3963-3976 (2000)

[12] S. Takahashi, M. Yoshiba and Y. Harada, "Nano-Characterization of Ceramic Top-coat/Metallic Bond-Coat Interface for Thermal Barrier Coating Systems by Plasma Spraying", Materials Transactions, **44**, 1181-1189 (2003)

[13] E.Berghof-Hasselächer, H.Echsler, P.Gawenda, M. Schorr and M. Schütze, Time and Temperature Dependent Development of Physical Defects in Thermal Barrier Coating Systems, Prackt Metallogr, **40**, 219-231 (2003)

[14] H. Echsler, D. Renusch and M. Schütze, Bond coat oxidation and its significance for life expectancy of thermal barrier coating systems, Materials Science and Technology, **20**, 307-318 (2004)

[15] M. Hasegawa and Y. Kagawa, "Microstructural and Mechanical Properties Changes of a NiCoCrAlY Bond Coat with Heat Exposure Time in Air Plasma-Sprayed Y2O3-ZrO2 TBC systems", International Journal of Applied Ceramic Technology, **3**, 293-301 (2006)

[16] Y. Yoshizawa, "New theory of error", 157-161 (1989) Kyoritsu-shuppan.

QUANTATIVE MICROSTRUCTURAL ANALYSIS OF THERMAL BARRIER COATINGS PRODUCED BY ELECTRON BEAM PHYSICAL VAPOR DEPOSITION

Matthew Kelly[1], Jogender Singh[1], Judith Todd[2], Steven Copley[1], Douglas Wolfe[1]
[1]The Applied Research Laboratory
[2]Engineering Sciences and Mechanics Department
Pennsylvania State University
University Park, Pa, 16802

ABSTRACT

Thermal Barrier Coatings (TBC) produced by Electron Beam Physical Vapor Deposition (EB-PVD) are used primarily for system critical components of power turbines. The performance of coatings is highly dependant on micro and nano structural features. This paper proposes and demonstrates quantitative microstructural analysis of TBC produced by the EB-PVD. Metallographic techniques were applied to surfaces parallel and perpendicular to the columnar growth direction. Multiple levels perpendicular to the columnar growth direction were prepared to identify descriptive statistical values for coating microstructure throughout coating thickness. Metallographic surfaces were imaged with Scanning Electron Microscopy (SEM) and evaluated with commercial image analysis software. Samples produced with different vapor incidence angles were evaluated by the proposed method and show distinct quantitative differences of column grain size, inter-columnar porosity, and levels of re-nucleation. Microstructural results will be presented as a function of coating thickness and vapor incidence angle.

INTRODUCTION

Thermal Barrier Coatings (TBC) produced by Electron Beam Physical Vapor Deposition (EB-PVD) are used primarily for system critical components of power turbines due to the increased coating life compared to thermal sprayed coatings and ability to coat components with active air cooling without sealing external cooling passages [1]. Under current and expected future operating temperatures up to 1400°C, failure of the coating will result in rapid metal structural component degradation, turbine performance decreases, and eventual system failure [2]. PVD coatings exhibit what has been described as columnar morphology, high aspect ratio grains oriented normal to the substrate separated by voids, which offer mechanical compliance parallel to the component surface [1, 3-5]. This structural compliance aids in reducing stress in the ceramic coating during thermal cycling and thickening of the Thermally Grown Oxide (TGO) due to high temperature oxidation when compared to fully dense structures [6].

Accurate, cost-effective characterization of coating structure is an essential tool for coating development and quality control. There are two main areas of material structure that should be evaluated: physical structure and atomic structure. The evaluation of atomic structure, which includes determination of crystallographic information and chemical composition, is handled well by bulk techniques such as X-Ray Diffraction (XRD) and Spectroscopy techniques as discussed elsewhere and is routinely reported [7, 8]. Evaluating and quantifying physical structure of Physical Vapor Deposited (PVD) coatings is often ignored in nearly all publications related to TBC due to the lack of reliable cost-effective techniques. The fact that physical structure is not commonly evaluated for PVD coatings is not surprising, considering the broad range of possible parameters needed to describe the complex range of geometric morphologies shown in classic Structural Zone Models (SMZ) [4]. However, this lack of information has created a void in relating structure to both process parameters and coating performance, causing many studies to relate machine specific parameters to testing based performance. While development based on the relation between machine specific process parameters and material performance is very beneficial to industrial scale development, it prohibits implementation of outside developed knowledge. Moreover, the lack of coating evaluation prevents accurate quality control of simple structural parameters such as grain size or porosity. Without quality control and standardization, coating performance can be unreliable and not ideal for the environmental use.

The physical structure of TBC varies throughout the thickness and is determined by the process parameters of the equipment used to produce it. Qualitatively it can be observed that columns are often thinner at the "root" or area closest to the substrate. Ballistic deposition models have explained that columns nucleate from initial particles, grow in size rapidly, compete during growth, merge, reach a maximum size for the given surface mobility, and eventually nucleation begins again on the deposited column surface [9, 10]. The rate at which changes take place and geometry of structure during the growth is process dependant. Additionally, three types of porosity can also be observed that coatings produced on substrates rotating in the vapor cloud. Most easily observed is the void space between columns (Type I inter-columnar porosity), which has been described to be ribbon-like and believed to be directly related to coating compliance [11, 12]. The second type of inter-columnar porosity has been described as the open "feathery" surface of columns (Type II intercolumnar porosity) and can be observed under high magnification of coatings produced under high rotation rates fractured through thickness [13]. Type II intercolumnar porosity is believed to affect the low temperature phonon thermal transmission [14]. Finally the third type of porosity is trapped voids on the sub-micron scale (intra-columnar) porosity, which reduces high temperature electro-magnetic thermal transfer [15-17].

These previously described structural changes with respect to thickness imply that neither a single technique nor position evaluation will be adequate in complete structural characterization. Useful measurements must include sets of measured parameters with respect to position during growth or detailed knowledge on growth dynamics of the material system. The following research summarizes and demonstrates metallographic techniques discussed previously on a variety of TBC produced by EB-PVD [18]. The described technique readily provides micron scale information on Type I inter-columnar porosity and columnar grain size with respect to coating thickness and is only a step in the direction of characterization of EB-PVD TBC physical structure.

EXPERIMENTAL
Sample Production
Samples were generated to evaluate the processing parameter effect of Vapor Incidence Angle (VIA) for an industrial scale prototype deposition system to simulate the range of structure seen on a complex geometry component similar to a turbine blade. Substrate material was nickel alloy 625 cut into 1.9 cm diameter buttons 0.476 cm thick. Buttons were polished smooth using successive steps of wet grinding with silicon carbide paper to an 800 grit finish. Substrates were "heat tinted" in air at 700°C for 15 minutes to form an oxide layer. The surface and the side of each test specimen were grit blasted in a Unihone brand grit blaster using high purity 400 micron-size aluminum oxide particles. The distance from the edge of the nozzle to the surface of the samples was approximately 38 cm, with a pressure of 200 kPa. The angle of the nozzle with

respect to the sample surface was 45° to minimize the amount of embedded Al_2O_3 particles incorporated into the substrate surface. The grit blast time on each sample varied between 10–15 seconds and was performed until a uniform matte finish was obtained. Grit blasted substrates were ultrasonically cleaned in acetone for 20 minutes, rinsed with methanol, ultrasonically clean again in methanol for 20 minutes, and dried with nitrogen gas.

Prepared substrates were then tack welded to strips of stainless steel foil that were tack welded to an 8.64 cm diameter mandrel having wedges milled at angles 0°, 15°, 30°, 45°, 60°, 75°, and 90° with respect to the primary vapor flux direction. The mandrel was loaded into an industrial prototype Sciaky Inc. EB-PVD unit consisting of six EB-guns and a three-ingot continuous feeding system described in previous papers [19]. The deposition chamber and gas feed lines were then evacuated to a pressure of 10^{-3} Pa. Two electron beams were then used on the graphite heater assembly to bring the heating surfaces to ~1200°C for 20 minutes before samples were positioned 28 cm above the center of a 4.93 cm diameter 7%wt Yttria Stabilized Zirconia (7YSZ) ingot lot number 1297726, provided by Trans Tech Inc. of Adamstown, MD. Samples were then set to rotate at 12 RPM and allowed to soak at an average temperature of 1020°C for 20 minutes. During the soak period oxygen was introduced near the samples at 100 sccm to grow a thin uniform oxide layer (TGO). Following the TGO period, samples were exposed to an established 7YSZ vapor at an average chamber pressure of 1.7×10^{-3} torr. Evaporation rate was established at 5.7 grams per minute, resulting in a deposition rate of 2 microns per minute. Deposition was continued for a period of 90 minutes, producing a 177± 4 micron-thick 7YSZ coating for the sample with normal incidence angle (0° VIA). Samples were left to cool under vacuum and 200 sccm oxygen flow for 10 minutes before venting to atmospheric conditions. Coatings exhibited what would be considered typical range of morphologies for TBC deposited at VIA in the 0° to 90° range and are shown in Figure 1.

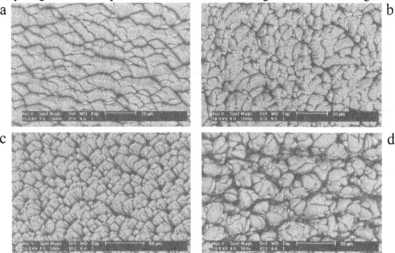

Figure 1. Typical range of morphologies shown a, b, c, and d for TBC deposited at Vapor Incidence Angles (VIA) of 0°, 30°, 60°, 90° respectively.

Sample Preparation

Samples were sectioned with a diamond waffering saw through the center of each disk along the line normal to the thickness variation, producing indistinguishable halves. One half was vacuum mounted in metallographic epoxy on a 50 micron sacrificial shim at taper inclination angles greater than 1°, yet less than 3° to minimize the magnification correction factor. Mounted samples were polished and imaged using Back Scattered Scanning Electron Microscopy (BSE) [18]. Image position with respect to Thermally Grown Oxide (TGO) was recorded as described by a previous article [18].

Image Analysis

Image analysis was conducted using commercial software produced by Clemex Technologies Inc. of Longueuil, Canada. Microstructural phase selection was determined using grey scale threshold limits. Voids were determined to have grey scale levels of the metallographic epoxy and lower. Porosity was measured as an area fraction based on the ratio of selected pixels to total image pixels. Grains were determined as areas with greater grey scale levels than the lightest fully developed grain boundaries. Fully developed boundaries are those that completely bisect column regions. Grains were separated using software built in functions that perform low level binary operations and verified by an operator for accuracy. The area of each separated grain was measured and reported as a diameter of a circle representing equivalent area.

RESULTS

Surfaces exposed using the described Transcolumnar metallographic method provided a surface that can be used to detail the growth of the coating through the entire thickness. Features such as intercolumnar porosity and columnar grain size, that are difficult to measure using the more common metallographic cross, section become very apparent with the Transcolumnar surfaces. Metallographic surfaces perpendicular to the growth direction eliminate the depth of view difficulties observed with conventional imaging techniques applied to cross sectional surfaces. These "Transcolumnar" surfaces provide additional contrast due to increased visible depth of pores and provide additional surface area available for evaluation at different positions with respect to distance from TGO. An example image of a thick coating displaying features of notable interest is shown in Figure 2. Figure 2a displays the most common measurable features visible with this evaluation technique; Type I inter-columnar porosity and columnar grains with defined boundaries. Highlighted in Figure 2b is Type II "feather like" inter-columnar porosity, developed re-nucleation, and co-competitive grain boundaries. As columnar re-nucleation becomes well defined by void like boundaries, it is classified as an independent grain. A select group of images with corresponding relative positions for Vapor Incidence Angles (VIA) 0°, 30°, 60°, and 90° are shown in Figure 3.

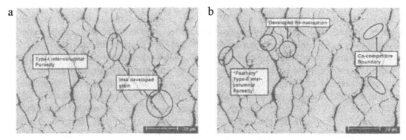

Figure 2. Shown in 2a are measured microstructural features of Type I inter-columnar porosity and developed columnar grains. Shown in 2b are notable features of observation; Region of Type II "feathery" inter-columnar porosity, region of developed re-nucleation, and a co-competitive grain boundary.

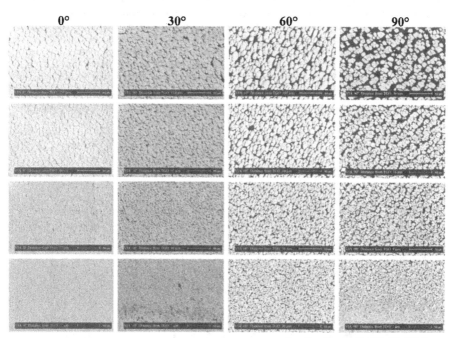

Figure 3. Sample images at consistent magnification for coatings produced at VIA of 0°, 30°, 60°, 90° shown in columns from left to right respectively. Position with respect to distance from TGO barrier in microns beginning with 0° VIA top image; 150, 96, 32, 11 micron. Position with respect to distance from TGO layer in microns for 30° VIA top image; 144, 95, 46, 7 micron. Position with respect to distance from TGO layer in microns for 60° VIA top image; 102, 69, 36, 20 micron. Position with respect to distance from TGO layer in microns for 90° VIA top image; 30, 16, 9, 3 micron.

Samples produced with different VIA were evaluated with the Transcolumnar metallographic method. It is clear that microstructural differences exist from visual inspection of SEM images shown in Figure 3. The most apparent visual difference is the changes of the void space between columns with both thickness and VIA. When quantitative image analysis was applied to the images for Type-I inter-columnar porosity and grain size the differences can be compared mathematically and graphically. Values of Type-I inter-columnar porosity are shown with respect to distance through thickness from the TGO layer graphically in Figure 4 for the different VIA. For the VIA 30° the porosity remains below 6% and somewhat constant with undulations. As VIA is increased above 30°, for this deposition process, porosity increases greatly. The amount of porosity becomes more dependent on thickness. Columnar grain size distributions were measured and the median grain size at each imaged position through the thickness is presented for the different VIA in Figure 5. Analysis of the mean grain size suggests that the minimum median grain size for all coating thickness under these deposition parameters should be found between 15° and 30° VIA. Additionally, the dependence of grain size on thickness suggests columns are growing divergently, or columnar grains are coalescing. Grain size distribution variance was plotted in Figure 6 to verify that re-nucleation and column divergence was the dominate trend. Parent-child image analysis could also be conducted with columns-columnar grains to make the same conclusion. One interesting observation when analyzing the grain size variance is that 15° VIA samples appear to have the tightest grain size distribution. Competitive growth may begin to dominate above 140 microns in thickness for these deposition parameters.

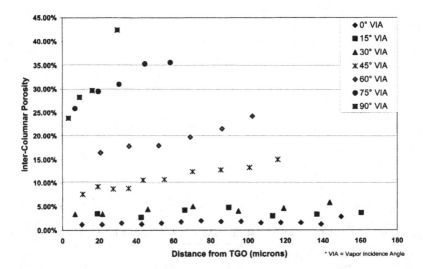

Figure 4. Measured values of Type-I inter-columnar porosity are shown with respect to distance through thickness from the TGO layer.

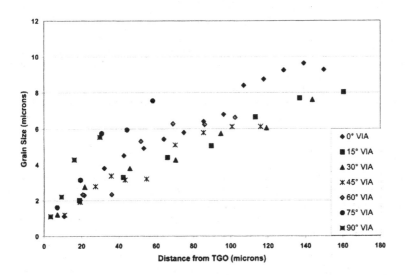

Figure 5. Measured values of median grain size are shown with respect to distance through thickness from the TGO layer.

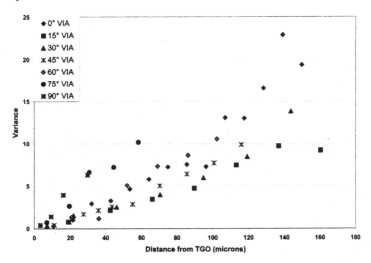

Figure 6. Calculated variance of grain size distribution is shown with respect to distance through thickness from the TGO layer.

Additional to the clear microstructural differences measured, trends on the growth mechanics can be observed. The most notable trend is the relation between grain size variance and porosity. Re-nucleation creates many small grains that grow rapidly, creating self-shadowing of nearby areas and increased porosity. This growth phenomenon can be observed by sharp changes in the slope of the graphs shown in Figures 4 and 6, and visually shown within in the images shown in Figure 3. This result agrees with accepted ballistic models for PVD and may offer insight into many of the unknown distribution functions needed for more complex growth modeling.

CONCLUSION

Thermal Barrier Coatings were applied to test coupons at discrete Vapor Incidence Angles on a rotating mandrill to simulate microstructures observed on complex components such as turbine blades. Samples were evaluated using a developed quantitative metallographic technique to evaluate microstructure at multiple levels through the thickness for Type-I inter-columnar porosity and columnar grain size. A ten fold increase of Type-I inter-columnar porosity was observed between normal incidence and samples parallel to vapor flux. Relative porosity and columnar grain size dependence was measured with respect to thickness. For the fixed process parameters, both porosity and columnar grain size were found to be dependant with coating thickness. Microstructural information corresponding to position with respect to thickness for a specified set of deposition parameters provides new knowledge into coating growth dynamics not obtainable through evaluation of historically evaluated surfaces. While research for this study was limited to the evaluation of Type-I inter-columnar porosity and columnar grain size distribution analysis, additional descriptive parameters can be measured from the same prepared samples.

Knowledge of microstructural information with respect to coating thickness for fixed deposition parameters is directly applicable for quality control. Exposed surfaces of deposited coatings can be imaged and analyized rapidly for columnar grain size and exposed porosity. Using predetermined charts as shown in figures 4-6 and measured thickness of the coating being evaluated for quality control purposes, an expected microstructure can be compared to the evaluated sample. Rapid verification of microstructure in this manner would provide an added level of assurance that coated components should perform as the tested samples during coating development.

Combinations of additional techniques and further development of metallographic microstructural evaluation techniques are required to achieve the level of resolution and microstructural information needed for true materials engineering of TBC coatings produced by EB-PVD. It is know that different microstructures of materials will perform differently under the same environmental usage. It was shown in this research that microstructure varies significantly with respect to geometry. Using the information retained from this study, it should be possible to design deposition parameters to increase microstructural uniformity on complex geometry substrates thereby increasing uniformity of coating performance. Incorporation of more advanced emerging techniques that accurately characterize the submicron levels of structure could provide the ability to design idealized material structures to maximize desired performance for individual applications.

ACKNOWLEDGEMENT

Distribution Statement A: Approved for public release. Distribution unlimited.
This research was sponsored by the United States Navy Manufacturing Technology (ManTech) Program, Office of Naval Research, under Navy Contract N00024-02-D-6604. Any opinions, findings, conclusions, or recommendations expressed in this material are those of the authors and do not necessarily reflect the views of the U.S. Navy.

REFERENCES

1. Peters, M., et al., *Design and Properties of Thermal Barrier Coatings for Advanced Turbine Engines.* Materialwissenschaft und Werkstofftechnik, 1997. **28**: p. 357-362.
2. Layne, A.W., *Advanced Turbine Systems*, N.E.T. Laboratory, Editor. 2000, U.S. Department of Energy.
3. Strangman, T.E., *Thermal barrier coatings for turbine airfoils.* Thin Solid Films, 1985. **127**(1-2): p. 93.
4. Thornton, J.A., *High Rate Thick Film Growth.* Annual Review Materials Science, 1977. **7**: p. 239-260.
5. Nicholls, J.R., M.J. Deakin, and D.S. Rickerby, *A comparison between the erosion behaviour of thermal spray and electron beam physical vapour deposition thermal barrier coatings.* Wear, 1999. **233-235**: p. 352-361.
6. Evans, A.G., et al., *Mechanisms controlling the durability of thermal barrier coatings.* Progress in Materials Science, 2001. **46**(5): p. 505.
7. Almeida, D.S., et al., *EB-PVD TBCs of zirconia co-doped with yttria and niobia, a microstructural investigation.* Surface and Coatings Technology, 2006. **200**(8): p. 2827.
8. Bernier, J.S., et al., *Crystallographic texture of EB-PVD TBCs deposited on stationary flat surfaces in a multiple ingot coating chamber as a function of chamber position.* Surface and Coatings Technology, 2003. **163-164**: p. 95.
9. Smith, D.L., *Thin-Film Deposition; Principles and Practice.* 1995: Mc Graw Hill. 616.
10. Hill, R.J., *Physical Vapor Deposition.* 1986: Temescal.
11. Clarke, D.R. and S.R. Phillpot, *Thermal barrier coating materials.* Materials Today, 2005. **8**(6): p. 22-29.
12. Terry, S.G., J.R. Litty, and C.G. Levi, *Evolution of Porosity and Texture in Thermal Barrier Coatings Grown by EB-PVD.* Elevated Temperature Coatings: Science and Technology III, 1999: p. 13-26.
13. Cho, J., et al., *A kinetic Monte Carlo simulation of film growth by physical vapor deposition on rotating substrates.* Materials Science and Engineering, 2005. **391**(1-2): p. 390-401.
14. Zhao, X., X. Wang, and P. Xiao, *Sintering and failure behaviour of EB-PVD thermal barrier coating after isothermal treatment.* Surface and Coatings Technology, 2005(Issues 20-21): p. 5946-5955.
15. Zhu, D., et al., *Thermal conductivity of EB-PVD thermal barrier coatings evaluated by a steady-state laser heat flux technique.* Surface and Coatings Technology, 2001. **138**(1): p. 1.

16.   Saruhan, B., et al., *Liquid-phase-infiltration of EB-PVD-TBCs with ageing inhibitor.* Journal of the European Ceramic Society, 2006. **26**(1-2): p. 49.
17.   Zhu, D. and R.A. Miller, *Thermophysical and Thermomechanical Properties of Thermal Barrier Coating Systems,* G.R.C. NASA, Editor. 2000, NASA Center for Aerospace Information. p. 22.
18.   Kelly, M.J., et al., *Metallographic techniques for evaluation of thermal barrier coatings produced by EB-PVD.* Materials Characterization, In Press.
19.   Wolfe, D.E., et al., *Tailored microstructure of EB-PVD 8YSZ thermal barrier coatings with low thermal conductivity and high thermal reflectivity for turbine applications.* Surface & Coatings Technology, 2005(190): p. 132-149.

INVESTIGATION OF DAMAGE PREDICTION OF THERMAL BARRIER COATING

Y. Ohtake
Ishikawajima-Harima Heavy Industries Co., Ltd.
1, Shin-Nakahara-Cho, Isogo-ku,
Yokohama-shi, Kanagawa 235-8501, Japan

ABSTRACT

Thermal barrier coating (top coating) for protecting turbine blades in airplane engines causes delaminalion by cyclic thermal loading. The delamination depends on the growth of thermal growth oxidation (TGO) layer at the interface when the coating system consists of top coating over environment barrier coating (bond coating). The growth behavior of TGO layer had examined by testing at constant temperature in furnace, but it didn't clear by testing of cyclic thermal loading. This paper investigates to examine the growth behavior of TGO layer for cyclic thermal loading. The burner rig testing is conducted to the behavior of TGO layer in cyclic thermal loading. Testing equipment is also developed to test four circular plate specimens at same time by burner rig. The thickness of TGO layer was measured from the observation at the interface in the specimen after the testing. The thickness increased as the number of cyclic thermal loading. The growth of TGO layer also denoted same tendency of the results of heating testing in furnace. It was found that the growth of TGO layer could predict by an equation. The equation was proposed in the relationships between the thickness of TGO layer and heating time when heating time was adopted for total holding times at maximum temperature in cyclic thermal loading.

INTRODUCTION

A typical coating system consists of thermal barrier coating (top coating) over environment barrier coating (bond coating). The coating system is applied for protecting engine parts, for example turbine blades, in airplane engines. The coating system fractures in top coating when the part is given cyclic thermal loading. The fracture mechanism is classified into two types. One is vertical crack in the normal direction of thickness of top coating. The cause depends on thermal expansion difference between top coating and base metal (or bond coating). Another is delamination of top coating at the interface over bond coating. The vertical crack can easily detect by regular inspection to appear on the surface of the engine parts, but the delamination can not detect in the first stage of the fracture because the crack propagates in top coating near the interface. Ohtake et al. had examined the damage of rectangular plate specimen with top coating by burner rig testing [1]-[4]. The delamination occurred in top coating from the observation of the specimen after the testing. The fracture mechanism was examined in both finite element analysis and burner rig testing. It was found that the delamination was caused by the growth of thermal growth oxidation (TGO) layer, thermal stress, the shape of interface at bond coating, the pore in top coating etc.

A simple life prediction model of top coating had been proposed from those results in previous paper [4]. The model is composed of three damage parameters. The growth of TGO layer is one of parameters in the model. The growth law of TGO layer had been examined in the results of the testing at constant temperature in furnace, but it didn't clear for cyclic thermal loading. This paper investigates to examine the growth behavior of TGO layer for cyclic thermal loading. The grown behavior of TGO layer was examined in burner rig testing. The testing takes in a lot of times even if it gives cyclic thermal loading to one specimen. The designer may also changes the material of base metal or the compositions (thickness, material and manufacturing process) of top coating. Thus, testing equipment was developed to test four circular plate specimens at same time by burner rig.

EXPERIMENTAL PROCEDURE

The specimen is composed of a typical coating system and base metal. The coating system consists of top coating over bond coating. Base metal is single crystal CMSX-2 substrate of nickel base superalloy. Bond coating applies CoNiCrAlY that is manufactured by low pressure plasma spray (LPPS). Top coating applies 8 wt. percent yttria stabilized zirconia (YSZ) that is manufactured by air plasma spray (APS).

Figure 1 shows the appearance in burner rig testing. The testing equipment was developed can test many specimens in short time. The equipment can test four specimens at same time by burner. Those specimens are circular plate and the size of the plate is diameter 20mm and thickness 3mm. Thickness of top coat of the specimen is 0.5m and bond coat 0.125mm. The surface of the specimen is heated by high temperature gas and the back surface is cooled by air. The thermal history of one cycle is total time 3 min, 20s heating time, 60s holding time and 100s cooling time. The maximum temperatures on surface of top coating are about 1473K, and then the temperature of the cooling surface are 1173K. The thermal loading is repeated until 1000, 2000 and 3000 cycles by burner rig. Those specimens are cut by diamond saw after the testing. TGO layer at interface in all specimens is observed by scanning electron microscope (SEM).

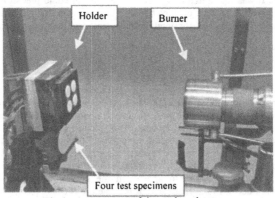

Fig.1 Appearance of thermal cycle test

The thickness of TGO layer is measured at ten points for one specimen and the average value is adopted for thickness of TGO layer of the specimen. Heating time is defined as the sum of holding time at maximum temperature in burner rig testing. The growth behaviors of TGO layer is examined from the relationship between thickness of TGO layer and heating time after cyclic thermal loading by burner rig testing.

EXPERIMENTAL RESULTS

Figure 2 shows the relationship between heating time and thickness of TGO layer, where heating time is total time at maximum temperature and the thickness of TGO layer is measured in the specimen at heating time in burner rig testing. TGO layer grows as the increase of heating time as shown in Fig.2 and the thickness increases as the number of cyclic thermal loading. The grown behaviors don't change for the testing at constant temperature in furnace. Thus, it was found that the grown of TGO layer didn't depend on testing method and heating time was important factor to predict the grown of TGO layer. Equation (1) was also proposed for the testing in furnace in previous paper [1]-[4]. The equation is expressed in terms of thickness of TGO layer w, heating time t and two constants k and n.

$$w = kt^n \qquad\qquad (1)$$

The line in Fig.2 is k=0.03 and n=0.3 in Eq.(1). The data of burner rig testing could predict by using the line of Eq.(1) in Fig.2. It was found that the growth of TGO layer in cyclic thermal loading could predict by Eq.(1).

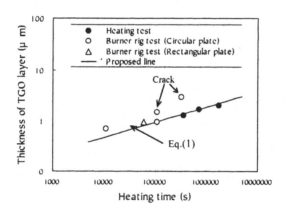

Fig. 2   Relationship between heating time and thickness of TGO layer

Crack was detected in top coat of the specimen after burner rig testing as shown in Fig.2 when the number of cycle reached more over 1000 cycles. The cracks existed in parallel direction of top coating surface from pore near the interface. It was considered that the delamination of top coating was occurred in the progress and the combination with a lot of cracks in top coating near the interface when TGO layer was increased with heating time.

CONCLUSIONS

This paper investigated growth behavior of TGO layer for cyclic thermal loading. The burner rig testing was used to examine the behavior of TGO layer in cyclic thermal loading. Testing equipment was also developed to test four circular plate specimens at same time by burner rig. The thickness of TGO layer after the testing increased as the number of cyclic thermal loading. The growth of TGO layer also denoted same tendency for the results of heating testing in furnace. It was found that the growth of TGO layer could predict by an equation. The equation was proposed in the relationships between the thickness of TGO layer and heating time when heating time was adopted for total holding times at maximum temperature in cyclic thermal loading.

REFERENCES

[1]Y. Ohtake, N. Nakamura, N. Suzumura and T. Natsumura, "Evaluation for Thermal Cycle Damage of Thermal Barrier Coating," *Ceramic Engineering and Science Proceedings*, 24(3) 561-566 (2003).

[2]Y. Ohtake, T. Natsumura, "Investigation of Thermal Fatigue Life of Thermal Barrier Coating," *Ceramic Engineering and Science Proceedings*, 25(4) 357-362 (2004). ·

[3]Y. Ohtake, T. Natsumura, K.Miyazawa, "Investigation of Thermal Fatigue Life Prediction of Thermal Barrier Coating," *Ceramic Engineering and Science Proceedings*, 26(3) 89-93 (2005).

[4]Y. Ohtake, "Damage Prediction of Thermal Barrier Coating," *Ceramic Engineering and Science Proceedings*, 27(3) (2006).

[5]R. A. Miller, "Oxidation-Based Model for Thermal Barrier Coating Life," Journal of the American Ceramic Society, 67 [8] 517-21 (1984).

[6]R. A. Miller, "Thermal Barrier Coatings for Aircraft Engines History and Directions," *Journal of Thermal Spray Technology*, 6 [1] 35-42 (1997).

[7]R. A. Miller, "Life Modeling of Thermal Barrier Coatings for Aircraft Gas Turbine Engines," *Journal of Engineering for Gas Turbines and Power*, 111 301-05 (1989).

[8]A. G. Evans, M. Y. He and J. W. Hutchinson, "Mechanics-based scaling laws for the durability of thermal barrier coatings," *Progress in Materials Science*, 46 249-271 (2001).

[9]A. G. Evans, D. R. Mumm, J. W. Hutchinson, G. H. Meier and F.S. Pettit, "Mechanisms controlling the durability of thermal barrier coating," *Progress in Materials Science*, 46 505-553 (2001).

[10]T. A. Cruse, S. E. Stewart and M. Ortiz, "Thermal Barrier Coating Life Prediction Model Development," Journal of Engineering for Gas Turbines and Power, 110 610-616 (1988).

[11]S. M. Meier, D. M. Nissley, K. D. Sheffler and T. A. Cruse, "Thermal Barrier Coating Life Prediction Model Development," *Journal of Engineering for Gas Turbines and Power*, 114 258-263 (1992).

Engineering of Thermal Properties of Thermal Barrier Coatings

EFFECT OF AN OPAQUE REFLECTING LAYER ON THE THERMAL BEHAVIOR OF A
THERMAL BARRIER COATING

Charles M. Spuckler
NASA Glenn Research Center
21000 Brookpark Rd.
Cleveland, OH 44145

ABSTRACT
    A parametric study using a two-flux approximation of the radiative transfer equation was
performed to examine the effects of an opaque reflective layer on the thermal behavior of a
typical semitransparent thermal barrier coating on an opaque substrate. Some ceramic materials
are semitransparent in the wavelength ranges where thermal radiation is important. Even with an
opaque layer on each side of the semitransparent thermal barrier coating, scattering and
absorption can have an effect on the heat transfer. In this work, a thermal barrier coating that is
semitransparent up to a wavelength of 5 micrometers is considered. Above 5 micrometers
wavelength, the thermal barrier coating is opaque. The absorption and scattering coefficient of
the thermal barrier was varied. The thermal behavior of the thermal barrier coating with an
opaque reflective layer is compared to a thermal barrier coating without the reflective layer. For
a thicker thermal barrier coating with lower convective loading, which would be typical of a
combustor liner, a reflective layer can significantly decrease the temperature in the thermal
barrier coating and substrate if the scattering is weak or moderate and for strong scattering if the
absorption is large. The layer without the reflective coating can be about as effective as the layer
with the reflective coating if the absorption is small and the scattering strong. For low
absorption, some temperatures in the thermal barrier coating system can be slightly higher with
the reflective layer. For a thin thermal barrier coating with high convective loading, which
would be typical of a blade or vane that sees the hot sections of the combustor, the reflective
layer is not as effective. The reflective layer reduces the surface temperature of the reflective
layer for all conditions considered. For weak and moderate scattering, the temperature of the
TBC-substrate interface is reduced but for strong scattering, the temperature of the substrate is
increased slightly.

INTRODUCTION
    Thermal barrier coatings (TBCs) are being developed for use in gas turbine engines.
TBCs can be made more effective by decreasing the heat conducted and/or radiated through
them. Some thermal barrier coatings are partially transparent to thermal radiation. For example,
for thermal radiation purposes zirconia can be semitransparent up to around 5 μm (refs. 1 and 2).
In semitransparent materials, both thermal radiation and heat conduction determine the
temperatures and the heat transferred. Scattering, absorption, emission, and the refractive index
determine the radiative heat transfer in a semitransparent material. The external and internal
reflection of an interface between two semitransparent materials depends on the refractive index
of the materials on each side of the interface. If thermal radiation is going from a material with a
higher refractive index to one with a lower refractive index, there is a total reflection of the
radiation at angles greater than the critical angle. Also, the thermal radiation emitted internally
by the material depends on the square of the refractive index. The internal thermal radiation
passing through the semitransparent interface is decreased by internal surface reflections, which

includes total internal reflection, so the energy emitted by the semitransparent layer can not exceed that of a blackbody. If there is an opaque layer on the semitransparent material the radiation emitted into the material depends on the refractive index squared and the emissivity of the opaque layer. The refractive index can have a considerable effect on the temperature profile in a semitransparent layer.

The scattering and absorption coefficients determine the amount of thermal radiation absorbed, emitted, and scattered by a semitransparent material. These coefficients have units of reciprocal length. The reciprocal of the coefficients can be considered as the mean distance traveled before absorption or scattering occurs (ref. 3 page 424). The smaller the coefficient the larger the distance thermal radiation will travel before being absorbed or scattered. When thermal radiation is absorbed or emitted by a material its temperature changes. Absorption and emission therefore have a direct effect on the temperature of a material. Scattered thermal radiation has no effect on the temperature of a material unless it is absorbed. Scattering in some cases can augment the absorption because it increases the path length of radiation through the material. Here scattering will act as additional absorption in determining the temperature profiles in a material ref. 4.

Putting a highly reflecting layer on the TBC is being considered as a method to improve their performance by reducing the radiative heat flux through the TBC. In reference 5 a gray semitransparent highly reflectance multilayered TBC system was designed and modeled. The results indicate the coating has potential for reducing the metal temperature. Here an opaque reflecting layer is considered. The absorption coefficient, scattering coefficient, and index of refraction still determine the radiative heat transfer through the semitransparent layer even though it is between opaque layers. Because scattering depends on the material structure and the absorption is affected by impurities and temperature, the absorption and scattering coefficients are increased and decreased from the base line values of $a = 0.1346$ cm^{-1} for the absorption coefficient and $\sigma_s = 94.38$ cm^{-1} for the scattering coefficient. These coefficients are a wavelength integrated average of those in ref. 2 for zirconia in wavelengths where it is semitransparent. The thermal behavior of a TBC with a highly reflecting opaque layer is compared to a normal TBC as a function of scattering and absorption.

MODEL

The models used, figure 1, are semi-infinite semitransparent layers on a substrate. One semitransparent layer does not have a reflective coating and the other has an opaque reflective coating. The semitransparent layer is semitransparent for thermal radiation up to 5 μm. For radiation above 5 μm the semitransparent layer is opaque. There is diffuse radiative and convective heat transfer on each side of the layers. The external radiative heating is q_{r1}° and q_{r2}°. The hot side gas and surrounding temperatures, T_{s1} and T_{g1}, cold side temperatures T_{s2} and T_{g2}, heat transfer coefficients, thickness of reflective coating, d_{rf}, TBC thickness d_{TBC}, and substrate thickness d_{sub} are given in Table 1. TBCs of the type that would be on a combustor liner and on a blade or vane are considered. For both the combustor and the blade and the emissivity of the back side of the metal substrate, ε_m, is 0.6. The thermal conductivity of TBC and the substrate are 0.8 w/mK and 33 w/mK. These conditions including those in Table 1 except for the reflective layer thickness were used by Siegel (ref. 6) to determine internal radiation effects in a zirconia based TBC. The thermal conductivity of the reflective coating, which is based on a dense zirconia-alumina multilayer coating, is 2.8 w/mK. The refractive index, n, of the semitransparent layer is 2.1. The refractive index of the gas is assumed to be one. The

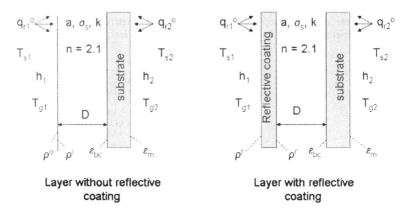

Layer without reflective coating   Layer with reflective coating

Figure 1 Heat transfer model

emissivity of the bond coat ε_{bc} is 0.3. The reflectance on both sides of the opaque highly reflective coating ρ^r is assumed to be 0.9. The external surface reflection for the layer without a reflective coating, ρ^o, was calculated using Fresnel's equation for a non-absorbing layer. This assumption should be good for the absorption coefficients used here (ref. 7 and ref. 3 page 88). The internal surface reflection, ρ^i, for the uncoated TBC was determined from a relationship

Table 1

	T_{s1} K	T_{g1} K	T_{s2} K	T_{g2} K	h_1 w/(m²·K)	h_2 w/(m²·K)	d_{rc} mm	d_{TBC} mm	d_{sub} mm
Combustor	2000	2000	800	800	250	110	0	1.0	0.794
Combustor	2000	2000	800	800	250	110	0.111	0.889	0.794
Blade	2000	2000	1000	1000	3014	3768	0	0.25	0.762
Blade	2000	2000	1000	1000	3014	3768	0.111	0.139	0.762

using the refractive index and external surface reflection in ref. 8. A two flux approximation to the radiative transfer equation was used to calculate the heat flux and the temperature profiles. The boundary conditions for the two flux equations in ref. 9 were modified to account for the opaque substrate and reflective coating.

EFFECT OF ABSORPTION AND SCATTERING ON COMBUSTOR LINER TBC WITH AND WITHOUT AN OPAQUE HIGHLY REFLECTIVE COATING

Temperature profiles

The temperature profiles in the TBC and substrate for a system with and without a highly reflective coating on a combustor liner as a function of absorption are shown in figure 2 for the base line scattering coefficient of 94.38 cm⁻¹. The temperature in the TBC and substrate decrease significantly when a 0.111 mm. opaque layer with a reflectivity of 0.9 is used. Depending on the absorption the decrease in the gas-TBC interface temperature is between about 127 K for 0.0013 cm⁻¹ absorption coefficient and 277 K for 67.26 cm⁻¹ absorption coefficient.

The TBC-substrate interface temperature decrease is approximately 173 K for 0.0013 cm^{-1} absorption coefficient and 213 K for 13.46 cm^{-1} absorption coefficient. The temperature

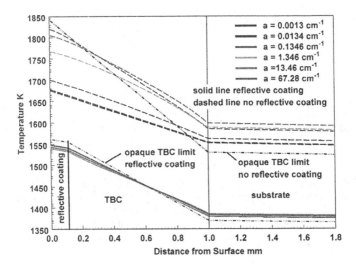

Figure 2. Temperature profiles in reflective coating, TBC, and combustor liner as a function of absorption for σ_s = 94.38 cm^{-1}.

difference due to a change in absorption also decreases significantly with a reflective coating compared to no reflective coating. Without a reflective coating the temperature change at the gas-TBC interface due to varying the absorption is about 145 K compared to 9 K with a coating. At the TBC-substrate interface, this temperature change is about 46 K without the reflective coating and about 6 K with the coating. Also, for this scattering coefficient, when there is no reflective coating the curvature of the temperature profiles indicate thermal radiation is playing a role. When an opaque reflective coating is present the thermal radiation role is decreased.

Interface temperatures and heat flux
        The effect of a highly reflective coating compared to no coating on the gas-TBC interface temperature as a function of absorption and scattering is presented in figure 3. The reflective coating reduces the gas-TBC interface temperatures significantly for higher absorption and all scattering considered. For lower absorption the effects of the reflective coating are reduced as the scattering is increased. For very low absorption and very strong scattering (high scattering coefficients) the gas-TBC interface temperatures are about the same for layers with and without a reflective coating. When a reflective coating is applied to a semitransparent layer the effects of absorption and scattering on the gas-TBC interface temperatures are reduced significantly compared to the uncoated TBC. Only for weaker scattering (smaller scattering coefficients) does absorption have an effect on the gas-TBC interface temperature when an opaque reflective

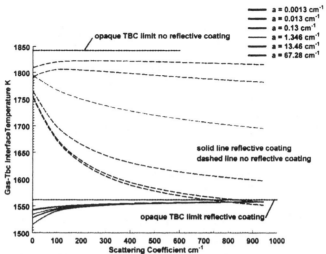

Figure 3. Gas-TBC interface temperature as a function of scattering and absorption for a TBC on a combustor liner

coating is present. The opaque limit shown in the figure is the temperature the TBC-gas interface would have if the TBC was opaque. For both the TBC without the reflective coating and the layer with the reflective coating the gas-TBC interface temperatures are less than the opaque limit, but for the TBC with a reflective coating the temperatures are approaching the opaque limit for strong scattering (higher scattering coefficients).

The temperatures of the TBC-substrate interface for two the layers are shown in figure 4. There is a large decrease in the TBC substrate interface temperature for weak scattering when a reflective coating is used. This temperature difference can be as high as 289 K. For very strong scattering and very low absorption this temperature difference decreases to around 10 K. The temperature difference decreases as scattering increases for all except the highest absorption and low scattering where the temperature difference decreases as absorption increases. When there is a reflective coating the effect of absorption and scattering is decreased substantially compared to a TBC without a reflective coating. For the layer with a reflective coating the effect of absorption decreases with scattering and only with weaker scattering is there an effect. For a TBC without a reflective coating the effect of absorption in general increases as the scattering increases. For a layer with an opaque reflective coating the TBC-substrate temperatures are higher for semitransparent TBC than an opaque TBC. For the TBC without the opaque reflective layer the TBC-substrate temperatures are lower than the opaque limit for some scattering and absorption coefficients. This was shown before in ref 10. The TBC-substrate interface temperature for a semitransparent TBC being higher than the opaque limit indicates an opaque TBC would perform better for these scattering and absorption conditions. The temperature profiles for the back of the substrate are similar to the TBC-Substrate temperatures and not presented here.

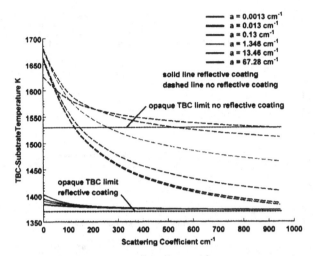

Figure 4 TBC-substrate interface temperatures as a function of scattering and absorption for a TBC on a combustor liner

The heat flux for layers with and without a highly reflective opaque layer as a function of absorption and scattering is shown in figure 5. The profiles and results are similar to the TBC -

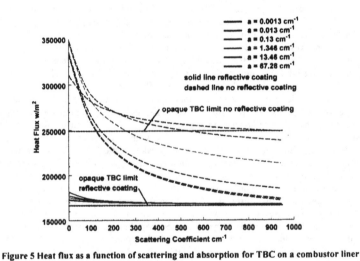

Figure 5 Heat flux as a function of scattering and absorption for TBC on a combustor liner

substrate temperatures. There is a significant decrease in the heat flux through the layer when a highly reflective coating is used especially for smaller scattering coefficients. The difference in heat flux decreases as the scattering coefficient is increased and decreases as the absorption coefficient is decreased except for the highest absorption coefficient and lower scattering coefficients. The effects of scattering and absorption are small and decrease as the scattering coefficient is increased when there is a reflective coating on the TBC. When there is no reflective coating on the TBC the effects of scattering and absorption can be large. With a reflective coating, the heat flux for a semitransparent TBC is larger than an opaque TBC with a reflective coating. This means it would be better to have an opaque TBC as far as the heat flux is concerned. With no reflective coating the heat flux for the semitransparent TBC is higher mainly for lower scattering coefficients. For the conditions where the heat flux is higher than the opaque limit, an opaque TBC would perform better.

EFFECT OF ABSORPTION AND SCATTERING ON BLADE OR VANE TBC WITH AND WITHOUT AN OPAQUE HIGHLY REFLECTIVE COATING

Temperature profiles

The temperature profiles for a TBC and substrate system for a vane or blade with and without a reflective coating are shown in figure 6. The thickness of the TBC is decreased from

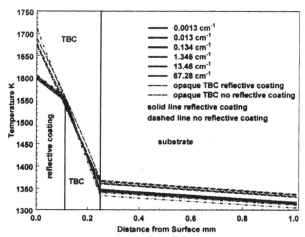

Figure 6 Temperature profiles in reflective coating, TBC, and blade or vane as a function of absorption for σ_s = 94.38 cm^{-1}.

that of a combustor and the convective loads are increased. Here the reflective coating is nearly half of the TBC thickness. The vane and blades are at the front of the turbine so they have a high radiation load for the combustor and soot. When a reflective coating is on the TBC, the surface temperature of the TBC for the base line scattering coefficient of 94.38 cm^{-1} is reduced between

70 and 94 K depending on the absorption. The temperature at the back of the reflective coating is about the same as that of a TBC with out a reflective coating at the same distance 0.111 mm. The TBC-substrate interface temperature decreases between 17 and 24 K depending on the absorption coefficient. The decrease in temperature at the back surface is between 15 and 22 K which is similar to the TBC-substrate interface temperature drop.

Interface temperatures and heat flux

The effect of scattering and absorption on the gas-TBC and gas-reflective coating

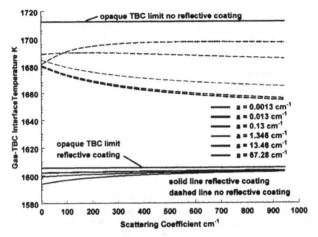

Figure 7. Gas-TBC interface temperature as a function of scattering and absorption for vane or blade

interface temperatures is shown in Figure 7. When there is no reflective coating, the temperature decreases with scattering except for the two highest absorptions coefficients used where the temperatures at first increases with scattering then decreases. The temperature increases with absorption except for the highest absorption used with weak scattering. The effects of absorption are small for absorption coefficients between 0.0013 cm^{-1} and 0.13 cm^{-1}. When there is a reflective coating the effect of absorption on the temperature is small and decreases as the scattering increases. For weak scattering the temperature variation due to absorption is about 12 K. For strong scattering there is essentially no temperature difference. For the TBC without a reflective coating and the TBC with the reflective coating the gas-surface interface temperature is lower than those that would occur if the TBC was opaque. The reduction in the gas TBC interface temperature with a reflective coating can be as high as about 97 K and as low as about 51 K depending on the scattering and absorption.

The temperature of the TBC-substrate interface as a function of scattering and absorption for a layer with and without a reflective coating are shown in Figure 8. The TBC-substrate interface temperature decreases with increased scattering and increases with increased absorption for a TBC with and without a reflective layer. The only exception is for a layer without a reflective coating where the temperature decreases for the highest absorption used when the scattering is moderate to weak. For weak scattering the TBC-substrate interface temperature is at most 40 K lower with a reflective coating. For strong scattering this reverses and the interface temperature is nearly 19 K lower without the reflective coating. For the TBC with the reflective layer the TBC-substrate interface temperature is slightly higher than the temperature that would occur if the TBC was opaque. For strong scattering and low absorption the temperature for the layer without a reflective coating is lower than the opaque limit. This indicates that for some conditions considered an opaque TBC without a reflective coating would perform better than a

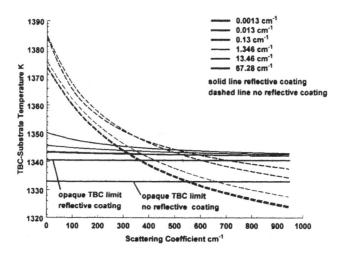

Figure 8 TBC-substrate interface temperatures as a function of scattering and absorption for a TBC on vane or blade

semitransparent TBC with or without a reflective coating if the thermal conductivity remained the same. It is also interesting to note that the temperature for an opaque TBC with a reflective coating is slightly higher that an opaque layer without a reflective coating. This is caused by the higher thermal conductivity coating taking up nearly half of the TBC thickness. Using a simple opaque analysis may give some insight on what thickness of a reflective coating can be tolerated. For the back of the substrate the results are similar to those for the TBC-substrate interface and are not presented.

The heat flux as a function of scattering and absorption is given in figure 9. The profiles and the results are similar to those for the TBC-substrate interface temperatures. For a TBC without a reflective coating the heat flux decreases as the scattering increases. The heat flux also increases as the absorption increases except the largest absorption coefficient when the heat flux decreases when the absorption is increased for weak to moderate scattering. When there is a reflective layer on the TBC the heat flux decreases as the scattering increases and in general increases as the absorption is increased. When there is a reflective coating, the effects of scattering on the heat flux are decreased significantly compared to a TBC without a reflective coating. Also the effects of absorption are reduced for strong scattering when a reflective coating is used. Here as with the TBC-substrate interface temperatures, the heat flux for a TBC with a reflective coating is decreased for weak to moderate scattering. For moderate to strong scattering depending on the absorption, the TBC without a reflective coating has a lower heat

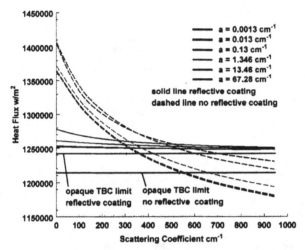

Figure 9. Heat flux as a function of scattering and absorption for TBC on a vane or blade

flux. For a TBC without a reflective coating with strong scattering and low to moderate absorption, the heat flux is lower than that of an opaque TBC. Here like the TBC- substrate temperatures, the heat flux for an opaque TBC with a reflective coating is higher than the heat flux for an uncoated opaque TBC. Also an opaque TBC will perform better than a semitransparent TBC over some of the conditions considered if the thermal conductivity remains the same. It seems like the scattering and absorption condition at which the change in performance occurs is the same for TBC-substrate temperatures and the heat fluxes.

SUMMARY AND CONCLUSIONS

A parametric study using a two-flux approximation to the radiative transfer equation was performed to examine the effects of an opaque reflective layer on the thermal behavior of a typical semitransparent thermal barrier coating on an opaque substrate as a function of scattering and absorption of the TBC. A one dimensional model was used. A TBC 1 mm thick on a 0.794 mm thick substrate with lower convection coefficients was used for the combustor TBC model. A 0.25 mm thick TBC on a 0.762 mm substrate with high convection coefficients was used to model the vane or blade TBC. The highly reflective coating was 0.111mm thick and this reduced the thickness of the underlying TBC to keep the total coating thickness unchanged. The substrate emissivity was 0.3 on the TBC side and 0.6 on the back side. The highly reflective coating has a reflectivity of 0.9 on both sides. The TBC was semitransparent up to 5 μm. Above 5 μm the TBC was opaque. There is radiative and convective heat transfer on each side of the layers. The absorption and scattering coefficients were varied.

For a combustor with a 1.0 mm thick TBC and convection coefficients 250 w/m^{2}K on the hot side and 110 w/m^{2}K on the cold side, the use of a 0.9 reflectivity coating 0.111 mm thick as part of the TBC can be effective in reducing the temperatures and heat flux in a TBC. The reflective coating is very effective at weak or low scattering at all absorption coefficients considered. For these conditions the gas-layer interface temperature decrease can be as high as about 280 K. For moderate to low absorption, the effectiveness of the uncoated semitransparent TBC increases with scattering so that adding a highly reflective coating does not add a significant benefit. For low absorption and strong scattering the gas-layer interface temperature can be higher with a reflective coating than without the reflective coating. At the TBC-substrate interface, the temperature decrease with a highly reflectivity coating can be as high as about 290 K for weak scattering depending on the absorption. The effectiveness of the reflective coating is reduced as the scattering is increased. For the lowest absorption and the highest scattering considered the temperature is 10 K lower when a reflective layer is used. The reduction in heat flux when a highly reflective coating is used is similar to the temperature reductions at the TBC-substrate interface. The largest decreases in heat flux occur for weak scattering. The reduction in heat flux with strong scattering can be quite significant if the absorption is large. It was also noted that when a reflective coating is put on a TBC the effects of the semitransparency, that is varying the scattering and absorption, are reduced drastically.

For a vane or blade with a 0.25 mm thick TBC, convection coefficients 3014 w/m^{2}K on the hot side and 3768 w/m^{2}K on the cold side, and a 0.9 reflectivity coating 0.111mm thick as part of the TBC, the reflective coating has mixed results on the temperatures and heat flux. The gas-TBC interface temperature is reduced by at least 50 K for all scattering and absorption considered. The largest difference in temperature in general occurs for high absorption. The TBC-substrate temperature deceases at least about 40 K for weak scattering. As the scattering increases the positive effects of the reflective coating decrease and become negative when the TBC-substrate temperature for the TBC with the reflective coating becomes greater than the TBC-substrate interface temperature for the uncoated TBC. The same phenomena occur for the heat flux. The decrease in effectiveness is probably due to the higher thermal conductivity reflective coating taking up almost half of the TBC thickness.

The TBC-substrate temperature and the heat flux for the combustor liner TBC without a reflective coating and the TBC–substrate temperatures and heat flux for a blade or vane TBC with and without a reflective layer were higher than that predicted for an opaque TBC material for some scattering and absorption conditions. For these conditions an opaque TBC with the same thermal conductivity would perform better than a semitransparent TBC.

REFERENCES
[1]Wahiduzzaman, S and Morel,T., Effect of Translucence of Engineering Ceramics on Heat Transfer in Diesel Engines, ORNL/Sub/88-22042/2, April 1992
[2]Makino, T., Kunitomo, T., Sakai, I., and Kinoshita, H., Thermal Radiation Properties of Ceramic Materials, *Heat Transfer-Japanese Research,* **13,** [4] 33-50 (1984)

[3]Siegel, R. and Howell, J. R. *Thermal Radiation Heat Transfer,* 4th ed. Taylor & Frances, New York, 2002

[4]Spuckler, C. M. and Siegel, R., "Refractive Index and Scattering Effects on Radiative Behavior of a Semitransparent Layer," *Journal of Thermophysics and Heat Transfer*, 7[2], 302-10 (1993)

5 Wang, D., Huang, X., and Patnaik, P. "Design and Modeling of Multiple Layered TBC System with High Reflectance," *Journal of Material Science,* **41**, [19] 6245-55 (2006)

[6]Siegel, R. "Internal Radiation Effects in Zirconia Thermal Barrier Coatings," *Journal of Thermophysics and Heat Transfer*, **10**[4], 707-9 (1996)

[7]Cox, R. L., "Fundamentals of Thermal Radiation in Ceramic Materials,"; pp. 83-101 in Symposium on Thermal Radiation of Solids, edited by S. Katzoff, NASA SP-55, 1965

[8]Richmond, J. C., "Relation of Emittance to Other Optical Properties," *Journal of Research of the National Bureau of Standards-C. Engineering and Instrumentation*, **67C** [3], 217-26 (1963)

[9]Siegel, R. and Spuckler, C. M., "Approximate Solution Methods for Spectral Radiative Transfer in High Refractive Index Layers," *International Journal of Heat and Mass Transfer*, **37** [Suppl. 1] 403-13 (1994)

[10]Spuckler, C. M., "Effect of Scattering on the Heat Transfer Behavior of a Typical Semitransparent TBC Material on a Substrate," Ceramic Engineering and Science Proceedings, **26**[3], 47-54 (2005)

OPTIMIZING OF THE REFLECTIVITY OF AIR PLASMA SPRAYED CERAMIC THERMAL BARRIER COATINGS

A. Stuke, R. Carius, J.-L. Marqués, G. Mauer, M. Schulte, D. Sebold, R. Vaßen, D. Stöver
Institut für Energieforschung (IEF)
Forschungszentrum Jülich GmbH
Jülich, Germany

ABSTRACT

For gas turbine applications, metallic components are coated with a ceramic thermal barrier coating (TBC). The mostly used ceramic material, yttria partially stabilized zirconia (YSZ), absorbs radiation of wavelengths below 5μm only weakly. However, it is in this wavelength range where most of the radiation by walls and gas is emitted within the gas turbine at service temperatures. The aim of this work is to optimize the diffuse reflectivity of the air plasma sprayed (APS) TBC by improving the coating microstructure such that it leads to an increase in the reflectivity of radiation and thus yields a more efficient thermal insulation of the underlying metallic substrate. Powder of different grain size distributions has been air plasma-sprayed under two different spray distances to produce different porosities. The transmission and reflection in the near infrared has been measured in an IR-spectrometer. Additionally, the absorption has been independently measured by means of the photothermal deflection spectroscopy (PDS). The influence on absorption by vacancies in the coating's as-sprayed state has been also investigated. By using the Kubelka-Munk two flux model, the scattering and absorption coefficient of the sprayed TBC corresponding to such model have been determined and correlated to the measured porosimetry. These two coefficients are used to estimate the stationary temperature distribution across the coating by solving numerically a two-flux model containing radiation.

INTRODUCTION

The increase in operating temperature in the gas turbines requires a better thermal insulation of the turbine's metallic components, which can be achieved by depositing on top of them a ceramic thermal barrier coating (TBC), usually made of yttria partially stabilized zirconia (YSZ). In the near infrared (IR) range of wavelengths below 5μm, YSZ is nearly transparent to radiation. Since it is within these wavelengths where most of the radiation in the combustion chamber is emitted, YSZ permits an important part of the energy flow to reach directly the underlying metallic substrate, being absorbed there. Hence a TBC optimization regarding radiation is necessary. Air plasma spraying (APS) as coating process offers the possibility of modifying the microstructure of the deposited TBC (pores and micro-cracks) in order to enhance the backscattering of radiation and thus better shield the metallic substrate.

This work is organized as follows: firstly the different sprayed powders and the used spraying parameters are discussed, together with the resulting TBC microstructures. The optical properties (reflectance, transmittance and absorbance) are investigated by means of two different techniques. The experimental results are then correlated to the Kubelka-Munk model in order to obtain an estimation for the absorption and scattering coefficients, which in the last section will serve to calculate the expected stationary temperature distribution within the coating system using the two flux model similar to the model of R. Siegel and C.M. Spuckler[1].

EXPERIMENTAL: POWDERS, TECHNIQUES AND COATINGS

The feed stocks were two 8wt% YSZ powders with different size distribution and structure: hollow spherical powder manufactured by spray drying (Sulzer Metco) and dense fused and crushed powder (Treibacher). The YSZ coatings were deposited on a steel substrate by atmospheric plasma spraying using a Triplex II gun (Sulzer Metco), with current 500A and a power of 57kW. The stand-off distance was chosen 150mm (in-house standard) and 300mm. The spray parameters and the resulting porosity of the coatings are given in Table I. Coatings from the spraying of hollow spherical powder at a spray distance of 150mm are defined as our standard. The last 3 samples where annealed in a furnace for 1 hour at 600°C.

Table I. TBC samples investigated

sample	injected powder with grain size distribution	spray distance	thickness	total porosity	heat treatment	optical measurement
05_09t	hollow sphere, spray dried $d_{10,50,90}$=7, 40, 77µm	300mm	390µm	22.1%	as-sprayed	IR
06_250t	hollow sphere, spray dried $d_{10,50,90}$=7, 40, 77µm	150mm	400µm	12.7%	as-sprayed	PDS
06_256t	hollow sphere, spray dried $d_{10,50,90}$=7, 40, 77µm	300mm	380µm	20.8%	as-sprayed	IR, PDS
06_508t	hollow sphere, spray dried $d_{10,50,90}$=12, 45, 88µm	150mm	390µm	12.0%	annealed 1h at 600°C	IR, PDS
06_509t	hollow sphere, spray dried $d_{10,50,90}$=12, 45, 88µm	300mm	350µm	19.9%	annealed 1h at 600°C	IR, PDS
06_510t	dense fused & crushed $d_{10,50,90}$=9, 20, 42µm	150mm	390µm	8.7%	annealed 1h at 600°C	IR, PDS

The in-flight velocity and temperature of the injected powder particles were measured during thermal spraying by means of the diagnostics system DPV2000 (Tecnar Automation, Canada), evaluating a total of 5000 particles at a location shortly before the molten particles impinge on the substrate to build the ceramic TBC. The temperature was measured by means of a two-color pyrometer (wavelengths: 787 and 995nm; temperature range: 2000–4000°C); the velocity was obtained using the time-of-flight method. For the hollow spherical powder, the shorter stand-off spray distance (150mm) leads to a larger particle velocity and temperature than for 300mm, since in the last case particles are already decelerating and cooling down. For the dense powder, also smaller in size, the velocity and temperature are significantly larger, due to a better injection and a longer stay within the plasma jet core.

The microstructures of the deposited YSZ-coatings were characterized using optical microscopy and scanning electronic microscopy (SEM). Both the pore distribution and the open porosity of free-standing coatings were determined by means of mercury intrusion porosimetry, with the results represented in Fig. 2; these porosimetry measurements are quite reproducible, as shown for samples sprayed under similar conditions. Polished cross sections with the microstructure of three representative YSZ-coatings (06_508t, 509t & 510t) are shown in Fig. 3. With increased spray distance, the porosity of the coating is clearly increased (both displayed in porosimetry and micrographs): this is a result of the decreased flattening of the impinging particles, since they are slower, as well as of the higher presence of non-molten particles. On the other hand, coating 06_510t shows a more dense structure, characterized by a reduced porosity; this agrees with the particle velocity measurement where a higher velocity for smaller particles leads to a higher flattening degree and a closer contact between deposited particles.

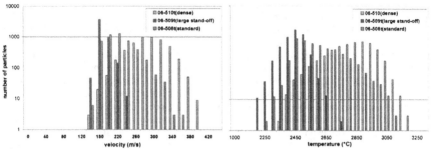

Figure 1: Particle velocity (left) and temperature (right) at impact for the hollow powder with standard (150mm) and large stand-off distance (300mm) as well as for the dense powder.

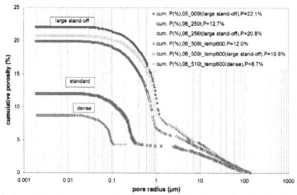

Figure 2. Double logarithmic representation of the cumulated pore distribution for the coatings.

Figure 3: SEM-micrographs of APS coatings 06_508t (hollow powder, in-house standard stand-off), 06_509t (hollow powder, large stand-off) and 06_510t (dense powder, standard stand-off).

For the optical characterization of the sprayed coatings (both as-sprayed as well as annealed) two different spectroscopic methods were used. The diffuse reflectance and transmittance measurements were performed by means of a Perkin Elmer UV-VIS-IR spectrometer, Lambda 950 in the wavelength range of 0.25 to 2.5μm. For collecting the scattered light, an integration sphere of 150mm diameter coated with TiO_2 was used.

Additionally, the absorbance of the samples were independently measured (0.5 to 2μm) using the photothermal deflection spectroscopy (PDS), particularly suitable for the determination of low absorbance ($\alpha*d=10^{-5}$).[2,3] Since this is a rather non conventional technique, a short description follows. The external surface of the sample to be measured is immersed in a liquid deflection medium (CCl_4). The sample is periodically excited (for instance by a laser modulated by a chopper). The absorption of the pump beam causes a periodic change of temperature at the sample surface, which for its part produces a gradient in refractive index in the deflection medium in contact to the surface (Fig. 4, left). The resulting periodic deflection of a laser passing along the sample surface is detected with position sensors (measuring both amplitude and phase). The deflection intensity is directly correlated to the optical absorption of the sample.

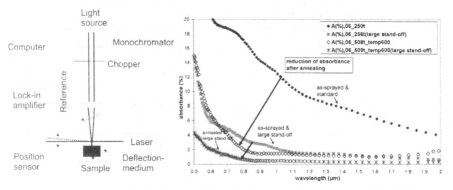

Figure 4: Left, schematic principle of the photothermal deflection spectroscopy (PDS). Right, influence of annealing on the absorbance of coatings sprayed at standard and large stand-off.

OPTICAL PROPERTIES OF THE SPRAYED COATINGS

The reflectance R and transmittance T for as-sprayed as well as annealed TBCs is represented in Fig. 5. Both properties display clearly the effect of the microstructure: with increasing porosity (from 8.7% to 19.9%) the reflectance increases and the transmittance decreases. In order to check the reproducibility of the measurements two samples (05-09t and 06-256t) sprayed with the same powder and stand-off were considered, showing a good agreement of two measurements. Further, similar results by Debout et al[1] has been included as reference. The sharp jump at 0.8μm is due to the change of the detector, and the strong data scattering for wavelengths above 2μm is due to the limited sensitivity of the IR detector. The superimposed periodic fluctuations in particular in the transmittance at 1.4, 1.9, 2.3μm are probably due to vibration modes of water-related species.

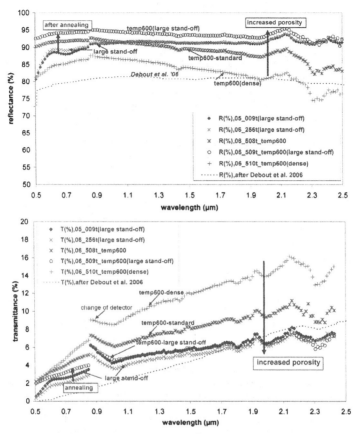

Figure 5. Reflectance (above) and transmittance (below) of as-sprayed and annealed coatings with standard and large spray distances measured with the IR integration sphere.

The coating 06_509t, annealed for 1h at 600°C, shows a higher reflectance than the corresponding as-sprayed coating 06_256t (the similar effect is not so clear in transmittance). This effect can be related to the sub-stoichiometry of oxygen, where a higher number of vacancies increase the absorbance of the coating, reducing thus the reflectance; this has been also discussed in Debout et al[4]. This interpretation is further supported when comparing the coating deposited at a shorter spray distance with that at larger stand-off: coating 06_508t, with a shorter stay in the plasma core region where atmospheric oxygen cannot diffuse inwards easily and therefore a higher deficiency in oxygen is expected, display a lower reflectance and a higher absorbance (the low transmittance is not able to point out this tendency clearly).

The previously discussed vacancies effect has been measured mainly in diffuse reflectance, which might not be sensitive enough since the considered coating's reflectance lies above 95%. The PDS method with its high sensibility at low absorption was consulted to confirm this effect. Comparing the IR and PDS spectra both methods provide the same trends. However, in the transition from VIS to IR, the absorbance A calculated from the IR integration sphere data (as $A=100-R-T$) lies above the absorbance directly measured with PDS (Fig. 6). The PDS measurement may lead to a slight under-estimation when applied on highly porous coatings: the open porosity will be filled with the fluid used for the deflection measurement. This reduced the reflectivity and may decrease the light pace resulting in an under-estimation of the absorbance. The effect of the oxygen deficiency is nevertheless confirmed in the PDS spectra (Fig. 4, right), both for standard as well as large stand-off distance.

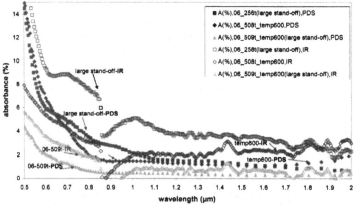

Figure 6. Absorbance of as-sprayed and annealed coatings with standard and large spray distances measured with PDS and compared to the IR integration sphere measurements.

ABSORPTION AND SCATTERING COEFFICIENTS OF THE SPRAYED COATINGS

For the following discussion, the annealed samples 06_508t, 509t and 510t have been selected in order to determine quantitatively the influence of the coating microstructure on the optical properties and on the estimated temperature distribution resulting from such optical properties. Only the annealed coatings correspond to the conditions in a gas turbine. The wavelength dependent absorption and scattering coefficients of the coating, $\kappa_{abs,\lambda}$ and $\kappa_{sca,\lambda}$, are obtained by fitting the IR integration sphere measurements to the Kubelka-Munk model for diffuse reflectivity and transmissivity of a coating. This model describes the 1-dimensional flow of radiation in forward and backward direction within an uniform system diffusely illuminated.[5] The absorption and scattering behavior of the system are effectively described by two parameters K and S which, assuming isotropic scattering, are related to the actual absorption and scattering coefficients by $\kappa_{ab,\lambda} = \frac{1}{2}K_{(\lambda)}$ and $\kappa_{xu,\lambda} = \frac{4}{3}S_{(\lambda)} + \frac{1}{6}K_{(\lambda)}$ (see Appendix B). The steps for obtaining K and S from the IR measurement of reflectance and transmittance are discussed in detail in the Appendix B. The results are represented in Fig. 7: taking as a reference the sample sprayed at the standard distance of 150mm with hollow powder (06_508t), the sample with the higher porosity (06_509t, sprayed at a larger stand-off distance) display as expected a higher scattering coefficient, whereas the dense sample (06_510t, sprayed with dense powder) has the highest absorption and the lowest scattering coefficients.

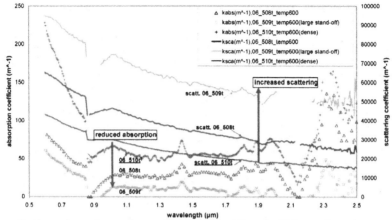

Figure 7. Absorption and scattering coefficients derived from the IR integration sphere measurements adjusted to the Kubelka-Munk model.

STATIONARY TEMPERATURE DISTRIBUTION IN SEMI-TRANSPARENT COATING

After having characterized the optical properties of the sprayed TBC coatings, the resulting stationary temperature distribution across the coating will be calculated in order to determine quantitatively the influence of the microstructure and its scattering ability. The model coupling the radiation transport within the semi-transparent coating to the temperature distribution has been developed by Siegel & Spuckler[1] and is discussed in some detail in Appendix A, together with its numerical solution. The system considered is formed by a ceramic TBC of thickness x_{TBC}=400μm (divided into 700 cells for the numerical solution) attached to a metallic substrate of thickness x_{SUB}=3.15mm.

The TBC is assumed to be semi-transparent for wavelengths $\lambda \leq \lambda_{cut}$=5μm (above this wavelength, the ceramic coating displays nearly black body behavior).[1] In order to smooth out the small jumps in the wavelength dependence of the absorption and scattering coefficients, the data of Fig. 8 between 0.6μm and 2μm are fitted to a power law $\kappa_{abs,scu,\lambda} = A\lambda^{-m}$ fit which is used to extrapolate both coefficients up to λ_{cut}=5μm. This extrapolation is justified by the fact that the slow decrease in the scattering coefficient should be maintained for longer wavelengths (it is the result of the pore distribution) and the absorption coefficient has already achieved almost saturation for wavelengths above 1.5μm.

The other model parameters are[1] (notation in Appendix B)

T_{gas}=2000K $\qquad$ $T_{sur,hot}$=1600K $\qquad$ $T_{sur,cold}$=800K $\qquad$ $T_{gas,cold}$=300K

ε_{gas}=0.22 $\qquad$ $\varepsilon_{surr,hot}$=1 $\qquad$ $\varepsilon_{surr,cold}$=0.6 $\qquad$ $\varepsilon_{gas,cold}$=0

$h_{conv,hot}$=250W/m²K $\quad$ n_{TBC}=n=2.1 $\qquad$ ε_{SUB}=0.3 $\qquad$ $h_{conv,cold}$=110W/m²K

with typical thermal conductivities λ_{TBC} =1.1W/mK for TBC and λ_{SUB}=26.2W/mK for the metal; the gas emissivity, based on the emission of CO_2 and H_2O, corresponds to a combustion chamber operating at a pressure of 10atm[6]. Within the TBC coating, the total energy flux is 0.231MW/m² for our standard spray distance (06_508t), 0.211MW/m² for the increased spray distance (06_509t) and 0.255MW/m² for the dense powder coating (06_510t). Compared to the our standard coating (06_508t), the sample with the higher porosity yields a lower temperature at the metallic substrate of about 36°C, whereas the sample dense sprayed has a temperature 7°C higher. In Fig. 8 the temperature distribution inside the TBC is represented, including a transparent TBC as comparison, and showing the better screening of radiation and the resulting lower substrate temperature for the coating of higher porosity. This reduces the thermal load on the metallic substrate and decreases the thermally activated growth of oxide scales at the interface ceramic-metal, leading to a higher life time of the coating system[8].

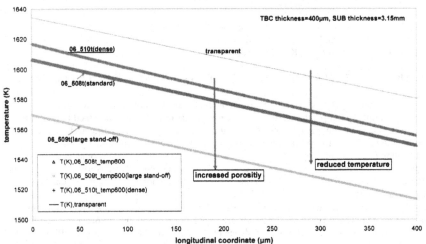

Figure 8. Calculated stationary temperature distribution across semi-transparent TBC (d=400µm) for the absorption and scattering coefficients of Fig. 7.

CONCLUSIONS

The IR optical properties of TBC coatings air plasma sprayed with different powders and stand-off distances have been investigated by means of two different optical techniques. The correlation to the microstructure as well as the effect of annealing was discussed and the absorption and scattering coefficients of the different coatings were determined. Finally these experimental results have been used to estimate numerically the stationary temperature distribution for the different considered TBC microstructures. The sample with the larger spray distance yields promising results for improving the shield of IR radiation by a modified APS TBC.

ACKNOWLEDGMENTS

The authors thank Mr F. Vondahlen and Mr. K.-H. Rauwald for the sample manufacture, Mr. M. Kappertz for the cross-section preparations and Mr. J. Klomfaß for the PDS measurements.

REFERENCES

[1]R. Siegel, C.M. Spuckler, "Analysis of thermal radiation effects on temperatures in turbine engine thermal barrier coatings," *Mater. Sci. Eng.* **A245**, 150-159 (1998).

[2]W.B. Jackson, N.M. Amer, A.C. Boccara, D. Fournier, *Applied Optics* **20** 1333 (1981).

[3]F. Becker, R. Carius, J.-T. Zettler, J. Klomfaß, "Photothermal Deflection Spectroscopy on Amorphous Semiconductor Heterojunctions and Determination of the Interface Defect Densities", *Materials Science Forum*, **173/174**, 177 (1995).

[4]V. Debout, A. Vardelle, P. Abélard, P. Fauchais, "Optical properties of Yttria-Stabilizied Zirconia Plasma-Sprayed Coatings" Proceedings of the *International Spray Conference ITSC2006*, Seattle, Washington (2006).

[5]R. Molenaar, J.J. ten Bosch, J.R. Zijp, "Determination of Kubelka-Munk scattering and absorption coefficients by diffuse illumination," *Applied Optics* **38**, 2068-2077 (1999).

[6]J.G. Knudsen *et al*, "Heat and Mass Transfer," section 5 of *Heat Transmission*, McGraw-Hill, New York (1997).

[7]M. F. Modest, *Radiative Heat Transfer*, Academic Press, Amsterdam (2003).

[8]F. Traeger, M. Ahrens, R. Vaßen, D. Stöver, "A Life Time Model for Ceramic Thermal Barrier Coatings," *Mater. Sci. Eng.* **A358**, 255-265 (2003).

APPENDIX A. TWO-FLUX RADIATION MODEL AND COUPLING TO STATIONARY TEMPERATURE DISTRIBUTION WITHIN SEMI-TRANSPARENT COATINGS

Let $I_\lambda(x;\hat{s})d\lambda$ be the radiation intensity at the location x along direction $\hat{s}$ and for wavelengths between λ and $\lambda+d\lambda$. The stationary change in intensity when moving to a neighbor location $x+dx$ is given by[7]

$$\cos\theta \frac{\partial I_\lambda(x;\hat{s})}{\partial x} = -\left(\kappa_{abs,\lambda} + \kappa_{sca,\lambda}\right)I_\lambda(x;\hat{s}) + \kappa_{abs,\lambda}n^2 I_{bb,\lambda}(x) + \kappa_{sca,\lambda} \underset{\text{whole sphere}}{\iint}\Phi(\hat{s},\hat{s}')I_\lambda(x;\hat{s}')d\Omega' \quad (A1)$$

with the cosine function for the projection of the intensity flow direction on the 1-dimensional coordinate x. The first term on the right hand side of (A1) describes the loss in intensity due to absorption and scattering, with respective spectral (i.e. wavelength dependent) coefficients $\kappa_{abs,\lambda}$ and $\kappa_{sca,\lambda}$. The second term carries the positive contribution due to the local black body emission $I_{bb,\lambda}(x)$ weighted by the absorption and thus emission coefficient at the considered wavelength

$\kappa_{abs,\lambda}$, with n the refractive index of the material in which the radiation is being transported. This term is considered since the considered coating usually operates at high temperatures and thus the emission of radiation by the system itself cannot be neglected. The last term in (A1) contains the intensity increase as a result from the scattering of intensities in all other directions $\hat{s}'$ back to the considered $\hat{s}$, where $\Phi(\hat{s},\hat{s}')$ represents the probability density for such processes fulfilling $\iint\limits_{\text{whole sphere}} \Phi(\hat{s},\hat{s}')d\Omega' = 1$.

In order to solve (A1) in a closed way, the following two simplifications are assumed: the scattering is assumed isotropic, such that $\Phi(\hat{s},\hat{s}')$ is a constant equal to $1/4\pi$. And the direction dependence in the intensity is reduced to only two main directions, forward and backward

$$I_\lambda(x;\hat{s}) \approx \begin{cases} I_\lambda^{(+)}(x) & \hat{s} \text{ in forward semi-sphere} \\ I_\lambda^{(-)}(x) & \hat{s} \text{ in backward semi-sphere} \end{cases} \tag{A2}$$

which is called the two-flux approximation. Now the number of unknowns at each location and wavelength has been reduced to $I_\lambda^{(+)}(x)$ and $I_\lambda^{(-)}(x)$ (or to their sum and difference), and thus two equations are required. The first one is obtained by integrating (A1) over the whole sphere

$$\pi\frac{\partial}{\partial x}\left(I_\lambda^{(+)} - I_\lambda^{(-)}\right) = -2\pi\left(\kappa_{abs,\lambda} + \kappa_{sca,\lambda}\right)\left(I_\lambda^{(+)} + I_\lambda^{(-)}\right) + 4\pi\kappa_{abs,\lambda}n^2 I_{bb,\lambda} + 2\pi\kappa_{sca,\lambda}\left(I_\lambda^{(+)} + I_\lambda^{(-)}\right)$$
$$= \kappa_{total,\lambda}(1-\omega_\lambda)\left[4\pi n^2 I_{bb,\lambda} - 2\pi\left(I_\lambda^{(+)} + I_\lambda^{(-)}\right)\right] \tag{A3}$$

with $\kappa_{total,\lambda} = \kappa_{abs,\lambda} + \kappa_{sca,\lambda}$ the (spectral) extinction coefficient and $\omega_\lambda = \dfrac{\kappa_{sca,\lambda}}{\kappa_{total,\lambda}}$ the scattering albedo. The second equation results from the multiplication of (A1) with $\cos\theta$ and integrating again over the whole sphere

$$\frac{2\pi}{3}\frac{\partial}{\partial x}\left(I_\lambda^{(+)} + I_\lambda^{(-)}\right) = -\pi\kappa_{total,\lambda}\left(I_\lambda^{(+)} - I_\lambda^{(-)}\right) \tag{A4}$$

By applying a further derivative on (A4), and together with (A3), it results

$$-\frac{1}{3\kappa_{total,\lambda}^2}\frac{\partial^2}{\partial x^2}\left[2\pi\left(I_\lambda^{(+)} + I_\lambda^{(-)}\right)\right] = (1-\omega_\lambda)\left[4\pi n^2 I_{bb,\lambda} - 2\pi\left(I_\lambda^{(+)} + I_\lambda^{(-)}\right)\right] \tag{A5}$$

which only contains the combination $G_\lambda(x) = 2\pi\left(I_\lambda^{(+)}(x) + I_\lambda^{(-)}(x)\right)$ as unknown. The energy flow due to radiation, equal to the integration of the radiation intensity over all directions, is related to the difference of $I_\lambda^{(+)}(x)$ and $I_\lambda^{(-)}(x)$ within the two-flux approximation

$$\dot{q}_{rad,\lambda}(x) = \iint_{\substack{whole\ sphere}} I_\lambda(x;\hat{s})d\Omega \approx \pi\left(I_\lambda^{(+)} - I_\lambda^{(-)}\right)^{eq.\ (A4)} = -\frac{1}{3\kappa_{total,\lambda}}\frac{\partial G_\lambda}{\partial x} \tag{A6}$$

Now let us consider a semi-transparent ceramic coating heated up at the external surface located at $x=0$ (corresponding to the combustion chamber) and attached on the other coating side at $x=x_{TBC}$ to a metallic substrate of thickness x_{SUB} which is cooled down. The external surface is heated due to convection and radiation by a flame at temperature T_{gas}, convection coefficient $h_{conv,hot}$ and emissivity ε_{gas} as well as due to radiation by the hot surrounding walls at temperature T_{sur} and emissivity ε_{sur}. The metallic substrate on the other side of the ceramic coating absorbs all the radiation impinging on it and has an emissivity ε_{sub}; on its free surface it interchanges energy to a cold gas flow at temperature $T_{gas,cold}$ (only convection, no radiation) and to the cold surrounding walls at temperature $T_{surr,cold}$ and emissivity $\varepsilon_{gas,cold}$ (radiation). The solution of the 2nd order differential equation (A5) for such a system requires two boundary conditions, for instance by giving the radiation flow $\dot{q}_{rad,\lambda}$ at the two coating boundaries[1]

$$-\frac{1}{3}\frac{1}{\kappa_{total,\lambda}}\frac{\partial G_\lambda}{\partial x}\bigg|_{x=0} = -\frac{(1-\rho_{int})}{2(1+\rho_{int})}\left[G_\lambda(x=0) - 4\pi\frac{1-\rho_{ext}}{1-\rho_{int}}\left(\varepsilon_{surr}I_{bb,\lambda}(T_{surr}) + \varepsilon_{gas}I_{bb,\lambda}(T_{gas})\right)\right]$$

$$-\frac{1}{3}\frac{1}{\kappa_{total,\lambda}}\frac{\partial G_\lambda}{\partial x}\bigg|_{x=x_{TBC}} = \frac{\varepsilon_{sub}}{2(2-\varepsilon_{sub})}\left[G_\lambda(x=x_{TBC}) - 4\pi n^2 I_{bb,\lambda}(T(x=x_{TBC}))\right] \tag{A7}$$

with the Fresnel diffuse reflectivities at the air-coating interface outwards as well as inwards[7]

$$\rho_{ext} = \frac{1}{2} + \frac{(3n_r+1)(n_r-1)}{6(n_r+1)^2} + \frac{n_r^2(n_r^2-1)^2}{(n_r^2+1)^3}\ln\left(\frac{n_r-1}{n_r+1}\right) - \frac{2n_r^3(n_r^2+2n_r-1)}{(n_r^2+1)(n_r^4-1)} + \frac{8n_r^4(n_r^4+1)}{(n_r^2+1)(n_r^4-1)^2}\ln n_r$$

$$\rho_{int} = 1 - \frac{1}{n_r^2}(1-\rho_{ext}) \tag{A8}$$

and with the relative refractive index at such interface $n_r = n/n_{air} = n$. Within the coating, the complete radiation energy flow integrated over the whole wavelength spectrum can be written as

$$\dot{q}_{rad}(x) = \int_0^\infty \dot{q}_{rad,\lambda}d\lambda = -\frac{1}{3}\frac{\partial}{\partial x}\int_0^\infty \frac{G_\lambda}{\kappa_{total,\lambda}}d\lambda = -\frac{1}{3}\frac{\partial}{\partial x}\int_0^{\lambda_1}\frac{G_\lambda}{\kappa_{total,\lambda}}d\lambda - \frac{1}{3}\frac{\partial}{\partial x}\int_{\lambda_1}^{\lambda_2}\frac{G_\lambda}{\kappa_{total,\lambda}}d\lambda - \dots - \frac{1}{3}\frac{\partial}{\partial x}\int_{\lambda_{M-1}}^{\lambda_{cut}}\frac{G_\lambda}{\kappa_{total,\lambda}}d\lambda$$

with λ_{cut} the largest wavelength at which the ceramic coating is still semi-transparent; for $\lambda > \lambda_{cut}$ the coating displays black body behavior and absorbs and emits immediately every wavelength perfectly, effect described by an infinite absorption (and thus extinction) coefficient. In the previous equation, the semi-transparent $[0,\lambda_{cut}]$ band has been divided in M windows $\{[0,\lambda_1], [\lambda_1,\lambda_2],\dots, [\lambda_{M-1},\lambda_M=\lambda_{cut}]\}$ within each one both the spectral extinction coefficient and the scattering albedo can be considered constant.

The total energy flow incoming on the coating consists of a convective part, due to the energy transfer from the hot gas flame, and a radiation part. The latter is divided into two contributions: that arising from the wavelength band where the coating is opaque such that its surface operates as a gray body of emissivity $1-\rho_{ext}$, absorbing the energy emitted by the gas and the walls in this band and radiating itself from the coating's heated surface (coordinate $x=0$); and the second contribution corresponding to the wavelength range where the coating is semi-transparent and therefore the radiation will be able to penetrate and propagate

$$\dot{q}_{total,in} = h_{conv,hot}\left(T_{gas,hot} - T\big|_{x=0}\right) + (1 - \rho_{ext})\pi \int_{\lambda_{wl}}^{\infty}\left(\varepsilon_{surr}I_{bb,\lambda}(T_{surr}) + \varepsilon_{gas}I_{bb,\lambda}(T_{gas}) - I_{bb,\lambda}\big|_{x=0}\right)d\lambda + \int_0^{\lambda_{wl}}\dot{q}_{rad,\lambda}d\lambda$$

Within the coating, the total energy flow consists of a conductive part as well as the just mentioned radiation able to propagate in the semi-transparent band, this time written as the separated contributions from each of the corresponding wavelength windows where the extinction coefficient is taken as constant

$$\dot{q}_{total,TBC} = \lambda_{TBC}\frac{T\big|_{x=0} - T\big|_{x=x_{TBC}}}{x_{TBC}} + \frac{1}{3}\int_0^{\lambda_1}\frac{G_\lambda\big|_{x=0} - G_\lambda\big|_{x=x_{TBC}}}{\kappa_{total,\lambda}}d\lambda + \ldots + \frac{1}{3}\int_{\lambda_{M-1}}^{\lambda_M}\frac{G_\lambda\big|_{x=0} - G_\lambda\big|_{x=x_{TBC}}}{\kappa_{total,\lambda}}d\lambda \quad (A9)$$

with λ_{TBC} the thermal conductivity of the ceramic coating. The metallic substrate absorbs all the incoming radiation and thus the total energy flow is $\dot{q}_{total,SUB} = \lambda_{SUB}\dfrac{T\big|_{x=x_{TBC}} - T_{metal,cold}}{x_{SUB}}$, i.e. only thermal conduction is relevant, with λ_{SUB} its thermal conductivity and $T_{metal,cold}$ the temperature at the externally cooled side. Finally, the total energy flow leaving the coating system through the cooled side consists of a convective part as well as the radiation as a gray body with emissivity ε_{sub} for the energy emitted by the metal surface together with the absorbed radiation emitted by the surrounding cold walls

$\dot{q}_{total,out} = h_{conv,cold}\left(T_{metal,cold} - T_{gas,cold}\right) + \varepsilon_{SUB}\sigma_{SB}\left(T^4_{metal,cold} - \varepsilon_{surr}T^4_{surr,cold}\right)$. In the stationary case, following holds

$$\dot{q}_{total,in} = \dot{q}_{total,TBC} = \dot{q}_{total,SUB} = \dot{q}_{total,out} \quad (A10)$$

which couples thus the radiation energy flow inside the TBC coating to the temperature distribution in the combustion chamber, in the metallic substrate and at the cooled side. This coupled equation system has to be solved numerically, discretizing the differential equation into a system of algebraic equations. For that the TBC coating thickness is divided into N cells of thickness $\Delta x[i]$ ($i=1,\ldots,N$), inside each one the temperature and energy flow is considered constant (although different from those in the neighbors cells). In the case of the radiation field variables, another index $[a]$ ($a=1,\ldots,M$) carries the information about the wavelength window within the semi-transparent band. The variables to be solved are

$$T(x) \to T[i], \quad \pi \int_{\lambda_{a-1}}^{\lambda_a} I_{bb,\lambda}(T(x))d\lambda \to \dot{q}_{bb}[a][i], \quad \int_{\lambda_{a-1}}^{\lambda_a} \frac{G_\lambda(x)}{\kappa_{total,\lambda}}d\lambda \to g[a][i]$$

(A11)

$$\int_{\lambda_{a-1}}^{\lambda_a} \dot{q}_{rad,\lambda}(x)d\lambda \to \dot{q}_{rad}[a][i] = -\frac{1}{3}\frac{g[a][i+1] - g[a][i-1]}{\Delta x[i+1]/2 + \Delta x[i] + \Delta x[i-1]/2}$$

And the discretization of equation (A5), incorporating the boundary conditions (A7), reads

$$\begin{pmatrix} b[1] & c[1] & 0 & \cdots & & \cdots & & \cdots \\ a[2] & b[2] & c[2] & 0 & \cdots & & \cdots \\ 0 & a[3] & b[3] & c[3] & 0 & & \cdots \\ \vdots & \vdots & \vdots & \vdots & \vdots & & \vdots \\ \cdots & \cdots & 0 & a[N-1] & b[N-1] & c[N-1] \\ \cdots & \cdots & & 0 & a[N] & b[N] \end{pmatrix} \begin{pmatrix} g[a][1] \\ g[a][2] \\ g[a][3] \\ \vdots \\ g[a][N-1] \\ g[a][N] \end{pmatrix} = \begin{pmatrix} r[1] \\ r[2] \\ r[3] \\ \vdots \\ r[N-1] \\ r[N] \end{pmatrix}$$

(A12)

$$a[i] = \begin{cases} 0 & i = 1 \\ -\dfrac{1}{3\Delta x[i]}\dfrac{1}{\Delta x[i]/2 + \Delta x[i-1]/2} & i \neq 1 \end{cases} \qquad c[i] = \begin{cases} 0 & i = N \\ -\dfrac{1}{3\Delta x[i]}\dfrac{1}{\Delta x[i+1]/2 + \Delta x[i]/2} & i \neq N \end{cases}$$

$$b[i] = -a[i] - c[i] + \kappa^2_{total,\lambda_a}(1-\omega_{\lambda_a}) + \begin{cases} \dfrac{\kappa_{total,\lambda_a}}{\Delta x[i]}\dfrac{(1-\rho_{int})}{2(1+\rho_{int})} & i = 1 \\ \dfrac{\kappa_{total,\lambda_a}}{\Delta x[i]}\dfrac{\varepsilon_{sub}}{2(2-\varepsilon_{sub})} & i = N \\ 0 & i \neq 1, N \end{cases}$$

$$r[i] = \kappa_{total,\lambda_a}(1-\omega_{\lambda_a})4n^2\dot{q}_{bb}[a][i] + \begin{cases} \dfrac{1}{\Delta x[i]}\dfrac{(1-\rho_{int})}{2(1+\rho_{int})}4\pi\dfrac{1-\rho_{ext}}{1-\rho_{int}}\int_{\lambda_{a-1}}^{\lambda_a}(\varepsilon_{surr}I_{bb,\lambda}(T_{surr}) + \varepsilon_{gas}I_{bb,\lambda}(T_{gas}))d\lambda & i = 1 \\ \dfrac{1}{\Delta x[i]}\dfrac{\varepsilon_{sub}}{2(2-\varepsilon_{sub})}4n^2\dot{q}_{bb}[a][i] & i = N \\ 0 & i \neq 1, N \end{cases}$$

Initially, for a guessed temperature distribution, equation system (A12) is solved for each wavelength window [a] in order to obtain the whole (provisional) distribution of g[a][i] through .the coating, for which the radiation flow is calculated from (A9). Subsequently, the new temperature distribution is obtained from (A10), which in its discretized form reads

$$T[i] = T[i=1] - \dot{q}_{total}\frac{x[i]}{\lambda_{TBC}} + \frac{1}{3\lambda_{TBC}}\sum_{a=1}^{a=M}(g[a][i=1] - g[a][i])$$

(A13)

Using this (new) current temperature distribution, the complete process is iteratively repeated until the change in the temperature distribution between two consecutive iterations is below 10^{-4}.

APPENDIX B. DETERMINATION OF THE ABSORPTION AND SCATTERING COEFFICIENTS FOR THE TWO-FLUX KUBELKA-MUNK MODEL

The two-flux Kubelka-Munk model is a simplification of the model in the previous Appendix, equations (A3) and (A4), without considering the radiation by the coating itself, since it is usually applied on systems operating at room temperature. Re-writing (A3) and (A4)

$$\frac{\partial I_\lambda^{(+)}}{\partial x} = -\left(\kappa_{abs,\lambda} + \frac{3}{4}\kappa_{total,\lambda}\right)I_\lambda^{(+)} + \left(\frac{3}{4}\kappa_{total,\lambda} - \kappa_{abs,\lambda}\right)I_\lambda^{(-)}$$

$$\frac{\partial I_\lambda^{(-)}}{\partial x} = +\left(\kappa_{abs,\lambda} + \frac{3}{4}\kappa_{total,\lambda}\right)I_\lambda^{(-)} - \left(\frac{3}{4}\kappa_{total,\lambda} - \kappa_{abs,\lambda}\right)I_\lambda^{(+)}$$

(B1)

which corresponds to the Kubelka-Munk model for the effective parameters K and S defined as
$K = 2\kappa_{abs,\lambda}$, $S = \frac{3}{4}\kappa_{total,\lambda} - \kappa_{abs,\lambda} = \frac{3}{4}\kappa_{sca,\lambda} - \frac{1}{4}\kappa_{abs,\lambda}$. Introducing $a = \frac{K+S}{S}$ and $b = \sqrt{a^2 - 1}$,
the reflectivity and transmissivity of a coating of thickness d solving (B1) are given by[7]

$$R_{KM} = \frac{\sinh(bSd)}{a\sinh(bSd) + b\cosh(bSd)}, \quad T_{KM} = \frac{b}{a\sinh(bSd) + b\cosh(bSd)}$$ (B2)

These equations can be considered as functions of the three independent variables d, b, $\zeta = bSd$. For a given thickness d the task now consists in searching for the combination of ζ and b that substituted in (B2) matches the measured reflectivity and transmissivity. For that it is worth noting that R_{KM} is a monotonous decreasing function of b and T_{KM} is monotonous decreasing in ζ. The search for ζ and b is carried out iteratively along the following steps:

1. Initializing the upper and lower limits of ζ: $\zeta_{up} = 1 \times d \times 10^6$ and $\zeta_{down} = 0$.

2. Begin of the iterative search for ζ: for the current $\zeta = \frac{\zeta_{up} + \zeta_{down}}{2}$, initializing the limits of the iterative search for b: $b_{up} = 1000$ and $b_{down} = 0$.

   a. With the current $b = \frac{b_{up} + b_{down}}{2}$, calculate $R_{KM} = \frac{\sinh\zeta}{\sqrt{b^2 + 1}\sinh\zeta + b\cosh\zeta}$ and
   $T_{KM} = \frac{b}{\sqrt{b^2 + 1}\sinh\zeta + b\cosh\zeta}$. Determine the next limits b_{down} and b_{up}: if R_{KM} is larger than the experimental reflectivity R then $b_{down} = b$ (R_{KM} is a decreasing function of b), otherwise $b_{up} = b$.

   b. Repeat the previous step until b_{down} and b_{up} are near enough to each other such that $2\frac{b_{up} - b_{down}}{b_{up} + b_{down}} < 10^{-4}$ holds.

3. Determine the next limits ζ_{up} and ζ_{down}: if the calculated T_{KM} is larger than the measured transmissivity T then $\zeta_{down} = \zeta$ (T_{KM} is a decreasing function of ζ), otherwise $\zeta_{up} = \zeta$.

4. Repeat steps 2 and 3 until $2\dfrac{\zeta_{up} - \zeta_{down}}{\zeta_{up} + \zeta_{down}} < 10^{-4}$ is fulfilled. Once finished, calculate

$$S = \frac{\zeta}{bd} \text{ and } K = S(a-1) = S\left(\sqrt{b^2 + 1} - 1\right).$$

THERMAL CONDUCTIVITY OF NANOPOROUS YSZ THERMAL BARRIER COATINGS FABRICATED BY EB-PVD

Byung-Koog Jang and Hideaki Matsubara
Materials Research and Development Laboratory,
Japan Fine Ceramics Center (JFCC)
2-4-1 Mutsuno, Atsuta-ku, Nagoya, 456-8587, Japan

ABSTRACT

ZrO_2-4mol% Y_2O_3 (YSZ) coatings were deposited by EB-PVD. The YSZ coatings consist of porous-columnar structure containing nano pores and intercolumnar gaps between columnar grains. The laser flash method and differential scanning calorimeter were used to measure the thermal diffusivity and specific heat of the coated samples. The thermal conductivity of EB-PVD coatings decreased with increasing measuring temperature as well as porosity. The thermal conductivity of the coating layer alone was also calculated based on thermal diffusion results for the double layers specimens consisting of the combined coatings and substrate. The response function method was employed to obtain an accurate value for the thermal conductivity of the coatings layer. The thermal conductivity of coatings layer showed increasing tendency with increasing the coatings thickness.

INTRODUCTION

Generally, for improvement of the thermal efficiency in gas turbine engine, higher operating temperature is necessary. However, metal component is always exposed at very severe operating temperature, resulting in the limit of metal substrate. In addition, a strong compressor is necessary for cooling which is negative for the thermal efficiency in the gas turbine engine. To overcome these handicaps, thermal barrier coatings (TBCs) have been developed for advanced gas turbine engine components to improve the thermal efficiency by increasing the gas turbine inlet temperature and reducing the amount of cooling air [1-3].

Currently, TBCs manufactured by electron beam-physical vapor deposition (EB-PVD) are being paid a great deal of attention because their columnar microstructure offers the advantage of a superior tolerance against thermal shock [4-6]. For superior TBCs, the development of coatings with the low thermal conductivity is very important. Therefore, thermal conductivity measurement of coatings is also necessary. Generally, it is known that the coatings have non-uniform porous structure so that they are easily damaged and broken due to the poor strength during measurement or handling. Actually, the evaluation of thermal properties of coatings is very critical issue.

The purpose of this work is to investigate the influence of temperature as well as porosity on thermal conductivity of EB-PVD ZrO_2-4mol% Y_2O_3 coatings by laser flash method. In addition, this work describes the thermal conductivity of coatings derived from heat diffusion results of the combined coatings and substrate specimens.

EXPERIMENTAL

ZrO_2-4mol% Y_2O_3 coatings were deposited by EB-PVD onto zirconia disk substrates of 10 mm diameter and 1mm thickness. The coating thickness was about 30~700 μm. The substrates were first preheated at 900~1000°C in a preheating vacuum chamber using graphite heating element. An electron beam evaporation process was conducted in a coating chamber using 30~60 KW of electron beam power. The target material was heated above its evaporation temperature of 3500°C. and the resulting vapors were condensed on substrate. Oxygen flow with 300 ccm/min could be fed into the coating chamber during deposition to control the stoichiometry of the YSZ coatings. Deposition was generally conducted in condition of 0~20 rpm of the substrate rotation for obtaining the different porosity of coatings. The substrate temperature was 950°C.

Free standing YSZ coatings layers were obtained by machining the substrate from the coated specimen prior to thermal conductivity measurements. The laser flash method is used to measure the thermal diffusivity of coated samples. The well-known laser flash method relies on the generation of a thermal pulse on one face of a thin sample and on the observation of the temperature history. The thermal diffusivity is determined from the time required to reach one-half of the peak temperature in resulting temperature rise curve for the rear surface as illustrated in Fig.1 [7].

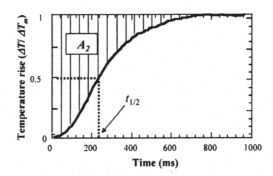

Figure 1. Temperature response behavior as a function of time at the rear surface of EB-PVD coated specimen after laser pulse heating.

All the measurements of the coated samples of 10 mm diameter were carried out between 25°C and 1000°C in 200 degree intervals in a vacuum chamber. Because of the translucency of the specimens to the laser, the specimens were sputter-coated with a thin layer of silver and colloidal graphite spraying to ensure complete and uniform absorption of the laser pulse prior to thermal diffusivity measurement. Specific heat measurements of the coated samples were made with differential scanning calorimeter (DSC) using sapphire as the reference material in an argon gas condition. The thermal conductivity of the coatings was then determined using

$$k = \alpha C \rho \qquad (1)$$

where k is the thermal conductivity, α is the thermal diffusivity, C is the specific heat, and ρ is the density of the coatings, respectively. The density of each coated sample was determined by measuring the mass and the volume of coated samples by a micrometer. The microstructure of the coated samples was observed by SEM. Raman spectroscopy was used to determine the crystal structures of the phases in the coatings.

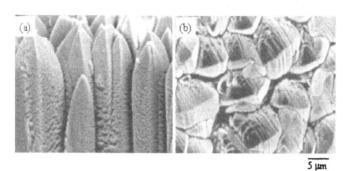

Figure 2. SEM micrographs of microstructure of ZrO_2-4 mol% Y_2O_3 coatings obtained by EB-PVD: (a) side view and (b) surface view.

RESULTS AND DISCUSSION

Typical microstructures in the coatings observed from surface and side regions of EB-PVD coatings are shown in Fig. 2. The top surfaces of the coatings consist of square-pyramidal or cone-like grains. In particular, the morphology of side regions of coatings deposited on substrates have a crystalline columnar texture with all columnar grains oriented in the same direction, namely perpendicular to the substrate, and with of a predominantly open porosity.

The columnar grains increase in size toward the top of column from the substrate, resulting in a tapered columnar structure. This result indicates that an epitaxial growth of YSZ films does not occur when deposited on substrates by EB-PVD. Gaps between columnar grains can also be clearly observed, particularly towards the top of the coatings. These gaps contribute to the porosity of EB-PVD coatings. The feather-like structure on both sides of the columnar grains contains many micro-sized, as well as nano-sized pores [8, 9].

Fig. 3 shows the result of Raman spectrum of the top surfaces of ZrO_2-4mol% Y_2O_3 coatings. This indicates that the observed phase in all coatings materials is the tetragonal phase of zirconia because its spectra were dominated by the relatively sharp tetragonal Raman modes.

Fig. 4 shows the porosity dependence on thermal conductivity for the free standing coatings layers separated from substrate. The porosity was calculated from the difference of density. The different porosity in the coatings was obtained by changing the rotation speed of substrate and multilayer between 300~700 μm.

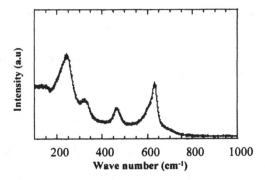

Figure 3. Raman pattern of ZrO_2-4 mol% Y_2O_3 coatings obtained by EB-PVD.

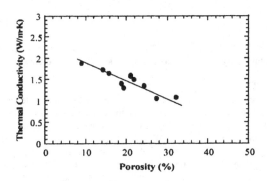

Figure 4. Relationship between thermal conductivity and porosity of
ZrO_2-4mol%Y_2O_3 coatings specimens obtained by EB-PVD.

The porosity showed decreasing tendency with increasing the rotation speed and multilayer.
It is seen that a wide variation in thermal conductivity is obtained in close relation with porosity.
The thermal conductivity of the free standing coatings layers decreases with increase of porosity.
Based on this result, it can be explained that the porosity is the dominant factor in determining
thermal conductivity for coatings by EB-PVD. This is consistent with results showing that the
porosity provides a major contribution to the reduction of the thermal conductivity of zirconia
coatings [10, 11].

Fig. 5 shows the results of thermal conductivity for free standing coated specimens from room
temperature to 1000°C. The thermal conductivity decreases slightly with increasing temperature
as well as porosity. It is readily apparent that the thermal conductivity of the coatings is well below

that of sintered ZrO$_2$-4mol% Y$_2$O$_3$ with full density.

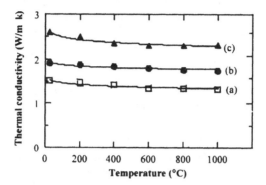

Figure 5. Thermal conductivity vs temperature of ZrO$_2$-4mol%Y$_2$O$_3$ specimens:
(a) coatings with 22% porosity, (b) coatings with 9% porosity obtained
by EB-PVD and (c) sintered bodies with full density.

For free standing specimens of very thinner coatings, the measurement of thermal conductivity is difficult because thinner specimens is easily broken or damaged during measuring or handling. For this reason, some researches on thermal conductivity with the laser flash method in the multi-layers system were reported [12,13].

When calculating the thermal diffusivity in the Fig.1, the method of determining $t_{1/2}$ values assumes that heat diffusion occurs across a uniform, pure and isotropic material. If this is the case, its value is reliable. However, the double layers specimens are non-uniform materials that consist of a porous coating layer and dense substrate similar to many multi-layer materials. Consequently, the method of estimating heat diffusion in the coated specimen to calculate the thermal conductivity of the coating layer must be reconsidered.

We suggested the calculation of thermal conductivity of coatings alone using the double layers specimens (coatings and substrate) using the response function method in the laser flash measurements [14].

Therefore, the thermal conductivity (λ_2) for coatings layer alone can be derived according to Eq. (2) using double layers specimens based on the response function method which the detailed theory was written in the previous work [7,15].

$$\lambda_2 = \frac{d_2^2 C_2 \rho_2 (3d_1 C_1 \rho_1 + d_2 C_2 \rho_2)}{6 A_2 (d_1 C_1 \rho_1 + d_2 C_2 \rho_2) - d_2^2 C_2 \rho_2 (d_1 C_1 \rho_1 + 3d_2 C_2 \rho_2)/\lambda_1} \tag{2}$$

where d_1, C_1, ρ_1 and d_2, C_2, ρ_2 correspond to the thickness, specific heat and density of substrate and coatings layer, respectively. λ_1 is the thermal conductivity of substrate. For double layers specimens, the area bounded by the temperature rise curve and the maximum temperature line at the rear face of the coated specimen after the laser pulse heating, designated A_2 in Fig. 1, can be obtained by integration. This area is called the "areal thermal diffusion time".

The areal thermal diffusion time as a function of coatings thickness in the present double layers specimens was given in Table 1.

Table 1. Calculated areal thermal diffusion time values as a function of coatings thickness for coated specimens.

Coating thickness (μm)	Areal thermal diffusion time (A_2)
79	0.208
119	0.223
249	0.279
297	0.298
302	0.305
337	0.324
368	0.333
497	0.415
613	0.489

Fig. 6 shows the correlation between the calculated and measured thermal conductivities for coatings layer alone with coating thickness based on Eq. (2). The calculated values of coatings show the good agreement with experimentally measured values. In the present results, the thermal conductivity of free standing coatings below 300 μm thickness cannot measure because of damage of specimens. The calculated thermal conductivity of coatings tends to increase with increasing of coatings thickness. The present results are consistent with the report of coating thickness on the thermal conductivity studied by Krell et al [16].

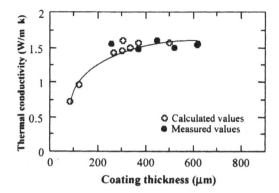

Figure 6. Thermal conductivity as a function of coatings thickness for
ZrO_2-4mol%Y_2O_3 coatings specimens obtained by EB-PVD.

In addition, the thicker coatings layer reveal higher thermal conductivity than that of thinner coated specimens. This reason can be considered that thinner coatings layer have very fine columnar grains and many columns boundaries without grain growth in column. Generally, phonons interact with imperfections such as dislocation, vacancy, pore and grain boundary. Therefore, many columns boundaries in thinner coating layers significantly make to reduce the mean free path by phonon scattering, resulting in the reduction of thermal conductivity [17].

The low thermal conductivity mechanism in the present porous coatings can be considered as following. The thermal conductivity can usually be reduced by decreasing the mean free path due to phonon scattering at pores. Many models exist to describe phonon heat conduction through porous media. The classical Maxwell model is the simplest model of thermal conduction for isolated pores dispersed in a continuous solid phase [18]. This model assumes an even distribution of random spherical pores and neglects pore geometry. Neglecting the heat conduction through the pores, this model can be expressed as:

$$k_{eff} \cong k_s \, [2(1-\phi)/(2+\phi)] \qquad (3)$$

where ϕ is porosity, k_{eff} is effective thermal conductivity of porous materials and k_s is thermal conductivity of the solid.

Fig. 7 shows the comparison of experimental thermal conductivity of the present specimens and theoretical thermal conductivity of porous materials based on Eq. (3). The theoretical thermal conductivity of the porous material including random spherical pores decreases with increasing porosity. The experimental thermal conductivity of the present specimens remarkably decreases with increasing porosity. The present specimens show lower thermal conductivity than theoretical values of the porous materials. The reason is that EB-PVD coatings consist of porous columnar grains with a feather-like structure containing evenly dispersed elongated pores as well as intracolumnar pores inside of the columns, resulting in the decrease of thermal conductivity due to the effective disturbance of heat flow.

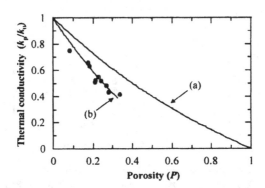

Figure 7. Normalized thermal conductivity as a function of porosity: (a) theoretical thermal conductivity and (b) experimental thermal conductivity of the present specimens.

CONCLUSIONS

ZrO$_2$-4 mol% Y$_2$O$_3$ coatings were deposited by EB-PVD. The coated layers had a columnar microstructure with intercolumnar gaps between columnar grains. Nano sized pores < 50 nm could be observed around feather-like grains as well as inside of columnar grains. The thermal conductivity of the coatings tended to decrease with increasing porosity. The thermal conductivity for coatings decreased slightly with increasing temperature between room temperature to 1000°C. The thermal conductivity of the coatings layer could be successfully calculated using the response function method applied to the combined coatings and substrate specimen. The calculated thermal conductivities of the coatings layer were in good agreement with experimental results from the laser flash method. Thermal conductivity of coatings increased with increasing coatings thickness. Increase of columns boundaries and pores in coatings layer mainly led to reduced thermal conductivity by the reduction of mean free path by phonon scattering.

ACKNOWLEDGMENTS

The authors acknowledge the financial support of the New Energy and Industrial Technology Development Organization (NEDO), Japan.

REFERENCES

[1] U. Schulz, B. Saruhan, K. Fritscher and C. Leyens, "Review on Advanced EB-PVD Ceramic Topcoats for TBC Applications," *Int. J. Appl. Ceram. Technol.*, **1**, 302-15 (2004).

[2] C. G. Levi, " Emerging Materials and Processes for Thermal Barrier Systems," *Current Opinion in Solid State and Mat. Sci.*, **8**, 77-91 (2004).

[3] D. D. Hass, P. A. Parrish and H. N. G. Wadley, "Electron Beam Directed Vapor Deposition of Thermal Barrier Coatings," *J. Vac. Sci. Technol.*, **16**, 3396-3401 (1998).

[4]J. Singh, D. E. Wolfe, R. A. Miller, J. I. Eldridge and D. M. Zhu, "Tailored Microstructure of Zirconia and Hafnia-based Thermal Barrier Coatings with Low Thermal Conductivity and High Hemispherical Reflectance by EB-PVD," *J. Mater. Sci.*, **39**, 1975-1985(2004).

[5]O. Unal, T. E. Mitchell and A. H. Heuer, "Microstructure of Y_2O_3-Stabilized ZrO_2 Electron Beam-Physical Vapor Deposition Coatings on Ni-Based Superallys," *J. Am. Ceram. Soc.*, 77, 984-92 (1994).

[6]T. J. Lu, C. G. Levi, H. N. G. Wadley and A. G. Evans, "Distributed Porosity as a Control Parameter for Oxide Thermal Barriers Made by Physical Vapor Deposition," *J. Am. Ceram. Soc.*, **84**, 2937-2046 (2001).

[7]B. K. Jang, M. Yoshiya, N. Yamaguchi and H. Matsubara, "Evaluation of Thermal Conductivity of Zirconia Coating Layers Deposited by EB-PVD," *J. Mater. Sci.*, **39**, 1823-1825 (2004).

[8]B. K. Jang and H. Matsubara, "Influence of Rotation Speed on Microstructure and Thermal Conductivity of Nano-Porous Zirconia Layers Fabricated by EB-PVD," *Scripta Mater.*, **52**, 553-558 (2005).

[9]B. Saruhan, P. Francois, K. Fritscher and U. Schulz, "EB-PVD Processing of Pyrochlore-Structured $La_2Zr_2O_7$-Based TBCs," *Surf. Coat. Technol.*, **182**, 175-183 (2004).

[10]K. An, K. S. Ravichandran, R. E. Dutton and S. L. Semiatin, "Microstructure, Texture, and Thermal Conductivity of Single-Layer and Multilayer Thermal Barrier Coatings of Y_2O_3-Stabilized ZrO_2 and Al_2O_3 Made by Physical Vapor Deposition," *J. Am. Ceram. Soc.*, **82**, 399-406 (1999).

[11]K. W. Schlichting, N. P. Padture and P. G. Klemens, "Thermal Conductivity of Dense and Porous Yttria-Stabilized Zirconia," *J. Mater. Sci.*, **36**, 3003-3010 (2001).

[12]R. F. Bulmer and R. Taylor, "Measurement by the Flash Method of Thermal Diffusivity in Two-Layer Composite Sample," *High Temp.-High Press.*, **6**, 491-497 (1974).

[13]J. Hartmann, O. Nilsson, J. Fricke, "Thermal Diffusivity Measurements on Two-Layered and Three-Layered Systems with the Laser-Flash Method," *High Temp.-High Press.*, **25**, 403-410 (1993).

[14] T. Baba, N. Taketoshi and A. Ono, "Analysis of Heat Diffusion Across Three Layer Thin Films by the Response Function Method" *21st Jpn. Symp. Thermophysical. Properties*, 229-231 (2000).

[15]B. K. Jang, and H. Matsubara, "Analysis of Thermal Conductivity and Thermal Diffusivity of EB-PVD Coating Materials," *Trans. of Mater. Res. Soc. Jpn.*, **29**, 417-420 (2004).

[16]H. J. R. Scheibe, U. Schulz and T. Krell, "The Effect of Coating Thickness on The Thermal Conductivity of EB-PVD PYSZ Thermal Barrier Coatings," *Surf. Coat. Technol.*, **200**, 5636-5644 (2006).

[17]J. R. Nicholls, K. J. Lawson, A. Johnstone and D. S. Rickerby, "Methods to Reduce the Thermal Conductivity of EB-PVD TBCs," *Surf. Coat. Technol.*, **151-152**, 383-391 (2002).

[18]H. Szelagowski, I. Arvanitidis and S. Seetharaman, "Effective Thermal Conductivity of Porous Strontium Oxide and Strontium Carbonate Samples," *J. Appl. Phys.*, **85**, 193-198 (1985).

Coatings to Resist Wear, Erosion, and Tribological Loadings

REDUCTION OF WEAR BY A TIBN MULTILAYER COATING

B.-A. Behrens; A. Küper; M. Bistron
Institute of Metal Forming and Metal-Forming Machines
An der Universität 2
Garbsen, 30823, Germany

Fr.-W. Bach; K. Möhwald; T.A. Deißer
Institute of Materials Science
An der Universität 2
Garbsen, 30823, Germany

ABSTRACT

Forging dies are exposed to high mechanical loads at the near surface area. Thermal and chemical stresses occur additionally. Depending on the number of forged parts different kinds of damage are developing in the surface area, which lead to failures of forging dies. Wear is with 70% the major cause of failure. In order to reduce wear, the abrasion resistance of the surface area of forging dies has to be increased. Therefore different methods of reducing wear on forging dies were examined, like the increase of the abrasion resistance by plasma nitrating and coating with ceramic layers (TiN, TiCN, TiC, CrN). These layers are applied on the surface by PACVD or PAPVD treatment. At the Institute of Metal Forming and Metal-Forming Machines a wear reduction by factor 3.5 compared to nitrided forging dies for forging of helical gears was achieved. This was possible by using a coating compound of 18 ceramic layers with an overall thickness of 1.8 µm. This paper deals with investigations of this multilayer compound and further research to reduce wear through an additional TiBN coating layer.

INTRODUCTION

High mechanical loads are applied on forging dies during the forging process. These loads superpose with thermal and tribological stresses in the near surface area. Depending on the number of forged parts these stresses cause different kinds of damage at the shaping area of the die[1]. As soon as the damage reaches a critical value it can cause the failure of a forging tool and the die has to be replaced. Thermal and mechanical crack formation, plastic deformation as well as wear are the main failure causes for the shaping area of forging dies (Fig. 1)[2]. Thereby the major occurring failure reason of a forging die is wear with 70%.

A significant reason for the occurrence of wear is the stability loss of the material in the near surface area of the forging die. This loss of stability is caused during the forging process by thermal stresses resulting from the high heat transfer from the hot billets and the comparatively cold surface of the forging die during of the pressure dwell time. The occurring temperature peaks can locally exceed annealing temperature of the tool material and cause a permanent loss in strength of the die material within the near surface area.

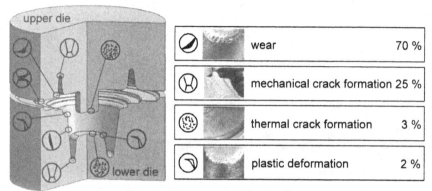

Figure 1. Failure causes of forging dies

Furthermore the shaping surface of the forging die is exposed to high, shock like temperature changes due to the use of cooling lubricants. This causes net-like cracks in the shaping area of the die which are related to the high temperature gradients. These cracks represent a starting point for abrasive wear. To increase the tool life time of conventional dies, which are made of hot working tool steel, the wear resistance of the die surface has to be increased. This effect can be achieved by nitriding the tools. Common nitriding techniques are salt bath- or gas nitriding. Recent research activities also consider plasma nitriding as a promising technique which gains more and more importance[3, 4, 5]. Another possibility to increase tool life time is a combination of plasma nitriding and hard coating. Thereby the nitrided layer becomes a supportive layer while the subsequently applied hard coating is the functional layer. Here special attention is paid to the increase of wear resistance of the near surface area by the use of hard material coatings.

EXPERIMENTAL SETUP AND FORGING TOOL

The forging tests at the Institute of Metal Forming and Metal-Forming Machines (IFUM) were carried out with a 3.15 MN eccentric press which is equipped with an automated billet handling suitable for long term tests. Constant and reproducible process parameters are guaranteed by additional equipment like a tool heating device and automated cooling and lubrication. The heating of the billet takes place in an inductive furnace. Sheared raw parts made from C45 (1.0503 / AISI 1043) are heated up to 1200°C for the forging tests. The average cycle time for each forged part is about 9 seconds. At investigations on a helical gear precision forging tool billets with a diameter of 30 mm and a height of 42 mm were used to forge without flash. Figure 2 shows the used forging tool and process parameters for the tests. After a given number of forging cycles the tool is dismounted for investigations. Metallographic analyses provide knowledge about microstructural changes of the tool material. To quantify the die wear the toothing of the die is investigated by a 3D measuring machine. For that purpose the tools are measured at first in unused condition to get a reference geometry for measurements of the die in different states of wear. By comparing the measured data of the worn-out tool with the reference geometry the progression of wear can be analysed in terms of the used process parameters.

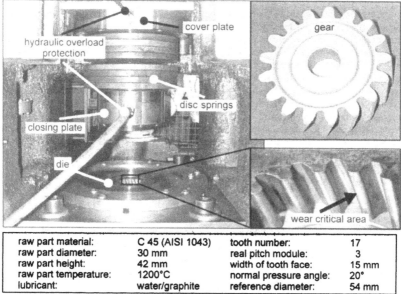

raw part material:	C 45 (AISI 1043)	tooth number:	17
raw part diameter:	30 mm	real pitch module:	3
raw part height:	42 mm	width of tooth face:	15 mm
raw part temperature:	1200°C	normal pressure angle:	20°
lubricant:	water/graphite	reference diameter:	54 mm

Figure 2. Helical gear precision forging tool

Previous investigations of a TiN-TiCN-TiC multi-layer coating system for this helical gear tool showed a reduction of wear by factor 3.5 compared to a nitrated forging tool after 2000 forging cycles[2, 6]. Based on this multilayer coating system a new coating system was developed with an additional TiBN top coating. This material was chosen to reinforce the die because of its high thermal and oxidation resistance at high temperatures. The left hand side of figure 3 shows a schematic drawing of the coating system.

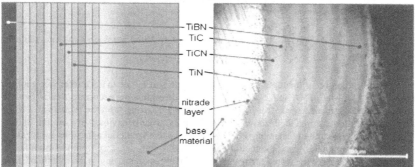

Figure 3. Schematic and microstructure (generated by a Calotest) of the multilayer coating

Plasma assisted chemical vapor deposition (PACVD) is used to apply the coatings in a continuous process on the hardened die. Before the coatings are applied the base material is plasmanitrided. By using adequate process parameters of the PACVD the brittle compound or so-called white layer can be prevented, because its material characteristics weaken the bond[7, 8, 9, 10]. After this a coating of 6-layers TiN-TiCN-TiC (18 single layers) and on top a final layer of TiBN is applied. Thereby the multilayer coating reaches an overall thickness of about 2.5 μm. Furthermore the right hand side of figure 3 shows a micrograph of the coating system applied on a specimen where the single layers are visible. Starting at the left hand side of the micrograph with the base material the changes of the coating from TiN to TiCN are visible. Between TiCN and TiC is no change of the material color noticeable because of the constantly increasing carbon ratio inside the coating material. After six layers of TiN-TiCN-TiC a clearly visible color change to TiBN can be seen at the right hand side of the micrograph.

MEASUREMENT AND RESULTS OF THE WEAR INVESTIGATIONS

For the measurement of wear a 3D measuring machine is used. Figure 4 shows how wear is measured and calculated. The profile and the top of three teeth are measured like depicted on the left hand side of figure 4. The right part of figure 4 shows how wear is calculated. A scan of the teeth shows the areas of material adhesion and material abrasion. Both deviations from the original geometry of the die are considered as wear because they affect the quality of the resulting work piece. The profile scans are made in the middle of the toothing due to the fact that the highest loads are expected there. By the use of the longitudinal measurement of the teeth the wear across the top of a tooth can be determined. Thereby it is possible to characterize the different loads during the forging process and its influence on the coating system.

Figure 4. Tooth measurement

The investigation of the toothing takes place at the unused die to get a reference geometry and then after 500, 1000, 1500 and 2000 forged parts to determine the occurring wear. The profile scans and the measured wear is depicted in figure 5. Starting from the profiles after 500 forged parts all measured teeth show a nearly identical wear behaviour. At the tooth tip is just little material abrasion noticeable. At both sides of the teeth material adhesions are located in the area between the tooth tip and the tooth flank. This adhesions decrease from the tooth flank to the tooth root. All three measured teeth show material abrasion starting in the middle of the right tooth flank. This is most likely a result of an inhomogeneous multilayer coating.

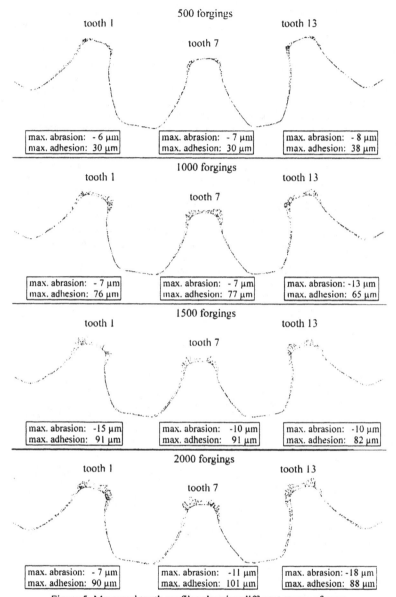

Figure 5. Measured tooth profiles showing different stages of wear

Additionally higher loads from the ejection of the gears compared to the left tooth flank aid the development of material abrasions in this area. At 1000 forgings the adhesions on the tooth tips and abrasions at the right tooth flank are still increasing. The location of the adhesions on the teeth is still nearly identical. Mostly the adhesions consist of scale, lubrication residues and abrasion from the forging die and billets[11, 12]. After the next 500 forged parts the adhesions at the tooth tips grow at a slower rate. Compared to the increase of adhesions from 0 to 500 and then up to 1000 forgings at an average rate of 38 μm an increase of material adhesions of about 15 μm can be found at forging up to 1500 parts. Additionally the tooth tips start to get a different shape. They all show areas of broken out adhesions and as it can be seen at the profile scan of tooth 1 after 1500 forgings even the coating is damaged by this break out of material (fig. 5). At tooth 13 the abrasions decreased about 3 μm during the last 500 forged parts. This and the fact that adhesions seem to grow at a slower rate lead to the assumption that the opposite is the case. The break out of material and deposition of new material nearly reached a balance. The profile scans after 2000 forgings confirm this effect. The increase of adhesions is still lower compared to the beginning but a closer look at the depicted profiles shows a very serrated shape of the tooth tips. The measurement of tooth 1 from 1500 to 2000 forgings shows a decrease of material abrasion and adhesion like seen at tooth 13 before.

As a conclusion of the depicted profile scans in figure 5, figure 6 shows the maximum resulting wear of the forging die. Additionally figure 6 shows that the adhesions after 1500 and 2000 forgings nearly stay at the same level. After 2000 forged parts there is still a reduction of wear compared to a nitrated forging die. With an average cycle time of 11 seconds the maximum wear value determined for a plasmanitrided die was 241 μm [12]. The exact wear reduction at a cycle time of 9 seconds needs to be verified but with the shorter cycle time and the resulting higher thermal loads the wear at a plasmanitrided die is most likely different.

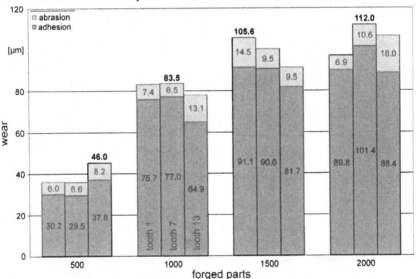

Figure 6. Calculated wear from the profile scans

To figure out at which point of the wear investigation the multilayer coating starts failing a closer look at the longitudinal scan of the tooth tip after 500 forgings is helpful. Figure 7 shows that the multilayer coating is already damaged in the early stages of the wear test. At least the 2.5 μm coating at the tooth tip is heavily damaged or destroyed.

500 forgings

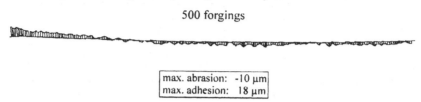

| max. abrasion: | -10 μm |
| max. adhesion: | 18 μm |

Figure 7. Longitudinal scan of the tooth tip

CONCLUSIONS

The objective of this paper is to present the performance of the TiN-TiCN-TiC multilayer coating system which is reinforced with a TiBN layer. This coating system was chosen because of its good results in wear protection. The intention of the final TiBN layer was to achieve an even better wear resistance performance through its high hardness and thermal resistance. After 2000 forgings a wear of 112 μm (101 μm adhesion and 11 μm abrasion) was measured by profile scans in the middle of a tooth. This means that with the first results of reinforcing the multilayer coating an additional increase of wear resistance could not be achieved. According to the profile scans a first failure of the coating is noticeable after 1500 forgings. Further measurements of the tooth tip show that the first failure of the reinforced multilayer coating appeared after 500 forgings already. The highest abrasion value is 10 μm. It is possible that the final TiBN layer prevented the flaking of the single coating layers and that starting cracks at the surface were piercing and weakening the multilayer coating. It is also possible that the hardness of the TiBN was to high compared to the last TiC layer and there were only particularly strong connections between the layers. This could lead to the flaking of the reinforcing layer and a rough surface which offers a starting point for wear. The exact reason for the early failure of the coating requires a closer investigation. Further wear tests with a TiB$_2$ reinforcement layer will possibly decrease the risk of layer flaking due to the high hardness difference between single layers. TiB$_2$ also offers a high hardness, good thermal stability and formation of less adhesions to decrease wear.

ACKNOWLEDGEMENT

The authors gratefully acknowledge the financial support of this work by the Deutsche Forschungsgemeinschaft (DFG). The goal of the Collaborative Research Centre 489 (SFB 489) is to develop new technologically and logistically innovative as well as economic process chains based on precision forging technology for the mass production of high performance components. This paper presents research results which were made within the particular project A1 "Precision Forging Materials" (Teilprojekt A1 „Werkstoffe für das Präzisionsschmieden").

REFERENCES

[1]T. Bobke, "Randschichtphänomene bei Verschleißvorgängen an Gesenkschmiedewerkzeugen," Dissertation, Universität Hannover, 1991

[2]A. Huskic, "Verschleißreduzierung an Schmiedegesenken durch Mehrlagenbeschichtung und keramische Einsätze," Dissertation, Universität Hannover, 2005

[3]W. K. Liliental; G. J. Tymowski; C. D. Morawski, "Typical Nitriding Faults and Their Preventation," Industrial Heating, 39-4 (1995)

[4]K. Ozasa; Y. Aoyagi, "Temporal evolution of hydrogen plasma with Gas pulse injection schem. Surface & Coatings Technology, 74-75, 345-350 (1995)

[5]K. S. Fancey; A. Leyland; D. Egerton; D. Torres; A. Matthews, "The influence of process gas characteristics on the properties of plasma nitrided steel," Surface & Coatings Technology, 76-77 694-699 (1995)

[6]B.-A. Behrens; A. Huskic; M. Gulde; A. Küper, "The Performance of Ceramic Coatings for the Use in Hot Massive Forming," The XVI International Scientific and Technological Conference "Design and Technology of Drawpieces and Die Stampings", Poland (2004)

[7]M. Van Stappen; M. Kerkhofs; C. Quaeyhaegens, "Introduction in industry of a duplex treatment consisting of plasma nitriding and PVD deposition of TiN," Surface and Coating Technology, 62, 655-661 (1993)

[8]T. Gredic; M. Zlatanovic; N. Popovic; Z. Bogdanov, Thin Solid Films, 228, 261 (1993)

[9]J. Walkowicz; J. Smolik; K. Miernik; J. Bujak, "Duplex surface treatment of moulds for pressure casting of aluminium," Surface and Coatings Technology, 97, 453 (1997)

[10]S. Walter, "Beitrag zu den Werkstoffversagensmechanismen beim Gesenkschmieden," Fortschritt-Berichte VDI, 5, 549 (1999)

[11]B. Bouaifi; A. Gebert, "Neuentwicklung und Verarbeitung von Schutzschichtwerkstoffen mit hoher Warmfestigkeit und Temperaturbeständigkeit. Development and processing of protective coating materials with high elevated temperature strength and temperature resistance," DVS-Berichte, 240, 158-163 (2006)

[12]B.-A. Behrens; L. Barnert; A. Huskic, "Alternative techniques to reduce die wear - Hard coating or ceramic? Alternative Techniken zur Verschleißminderung - Hartstoffüberzüge oder Keramik?," Production Engineering. Research and Development, 12-2, 131-136 (2005)

CHARACTERISTICS OF TiN/CrN MULTILAYER COATINGS WITH TiCrN AND CrTiN INTERLAYER

Xingbo Liu, Chengming Li, Jing Xu
West Virginia University
Morgantown, WV USA

Weizhong Tang, Fanxiu Lv
University of Science & Technology Beijing
Beijing China

ABSTRACT

Nanolayered TiN/CrN multilayer coatings with TiCrN and CrTiN interlayer were deposited on high speed steel M2 substrates by linear filtered arc ion plating (LFAIP). The modulation period of nano-multilayer coatings varies from 11nm to 400nm. The range of higher hardness modulation period is narrow, generally, from 5nm to 20nm. Because alloying nitrides formed interface between each single layer, modulation period of high hardness are extended from low modulation period area to high modulation period area in this paper designed multilayer system. Nanoindentation, scratch test, scanning electronic microscopy (SEM), X-ray diffraction (XRD), and three-point bend tests were used to characterize the coatings. The maximum hardness of the TiN/CrN multilayer with TiCrN and CrTiN interlayer was 33.2GPa. When modulation period is more than 70nm, the hardness of the TiN/CrN multilayer is reversely proportional to the square root of modulation period, which is consistent with Hall-Petch equation. However, it was observed a deviation from the linear Hall-Petch relation, when the modulation period is less than 70nm. The results of scratch test show the adhesion of TiN/CrN multilayer with TiCrN and CrTiN interlayer up to 80N. The load of cracks appeared on the surface of coatings at 60N and discontinuous chipping can be seen when the load reaches to 80N. The fracture morphology and cross-sectional of the coatings displayed the multilayered character. The deflection of crack propagation and micro-pinhole closed in films were observed due to multilayered formation. The emergence of small particles (<2 microns) still can be examined in the coatings, though magnetic filters and substrate pulse bias were used.

INTRODUCTION

Nano-multilayer nitride films has aroused great interest in recent years since it allows deposition of denser and more compact coatings, leading to improved chemical and mechanical properties [1].The Nitride nano-multilayer materials has emerged that exhibit extremely high hardness, making them very attractive for tribological applications. Multilayer consisting of very thin (5-20 nm) layers of nitride materials were deposited by filtered arc and magnetron sputtering, and these coatings have hardness in excess of 5000 kg/mm^2 [2]. This hardness is comparable to that of cubic-BN and is second only to diamond. For cutting tool applications one not only needs to prepare hard and tough coatings of plane but conformability of thickness and heterogeneous structure of their edge and corner are also very important.

TiN/CrN multilayer system is technologically important since it is expected to have good oxidation and corrosion resistance, and the superior mechanical properties, as compared to other nitride systems. The properties of TiN/CrN multilayer system have been studied in the literatures [3-10]. It was found that the modulation period Λ (sometimes called modulation wavelength or bilayer period), is the most important parameter of the coating, since the high hardness generally occurs in a narrow range of Λ of $\approx$ 5-20 nm. In industrial sized coating equipment of linear filtered arc ion plating (LFAIP), however, it is very difficulty to ensure conformability of thickness in nano-size grade range in the case of sharp corners and edges because modulation period occur changes due to significant changes of the plasma sheath surrounding a sharp edge [11-13]. Therefore, the small thickness of the deposited film at the cutting wedge region significantly affects the cutting performance [14]. The objectives of this paper are to study extending modulation period range of high hardness in TiN/CrN multilayer system with TiCrN and CrTiN interlayer and to investigate mechanical behavior and cutting performance of these new films.

EXPERIMENTAL
Film Deposition
A cathodic vacuum arc deposition system with linear magnetic filters was used to deposit the multilayer coatings. The system comprises four vacuum arc sources situated at 90°to each other, each capable of producing depositing materials by evaporation and magnetic filtering. High purity Ti and Cr were used as targets. High speed steel M2 and stainless steel 304 were selected as substrate materials. Each substrate and sample was polished, ultrasonically cleaned, degreased and then rinsed in deionized water, alkali and acetone baths, before they were mounted onto rotating fixture. The base pressure of the vacuum chamber was 8×10^{-4} Pa and its leak rate was lower than 5×10^{-2} Pa s^{-1}. The reactive gas used was high purity nitrogen and the working gas used to sputter the substrates and to ignite the arcs was high purity argon. Nitrogen flow rate was controlled via a proportional integrating differential (PID) controller, while the argon flow rate was kept constant. The pulse negative bias voltage with adaptable duty ratio was applied to the substrate. The substrate and samples were subjected to Ar$^+$ bombardment at −1000V pulsed bias to ensure good adhesion of the depositing coatings. To deposit TiCrN interlayer, the arc discharge on surface of Ti and Cr target are occurred simultaneously. After optimizing the deposition conditions for TiN/CrN multilayer coatings with TiCrN and CrTiN interlayer were prepared at different modulation wavelengths. Substrate temperature and substrate biases were kept at 450°C and -150V respectively.

Film Characterization

X-ray diffraction (XRD) was used to investigate structure and growth orientation of multilayer films. Coating adhesion was characterized with a conventional scratch tester. Three-point bend test was used to examine the toughness and crack propagation of the films. The hardness measurements were performed by means of a nanoindenter at a load of 5mN. At this load the indentation depth was much less than 1/10th of the coating thickness, thus eliminating the effect of substrate on the hardness measurements. Ten indentations were made on each sample and results presented herein represent the averages of 10 indentation. The cross-section morphologies of the coating including crack propagation and layered characteristics were characterized by scanning electronic microscope (SEM).

RESULTS AND DISCUSSION
X-ray Diffraction of Coatings

X-ray diffraction has been used to characteristic qualitatively for multilayer structure. High-angle X-ray diffraction patterns are generally taken in the θ-2θ geometry, in which the scattering vector is perpendicular to the interfaces in the multilayer. The high-angle XRD patterns consist of a convolution of the lattice spacing variation and the composition modulation. However, they also contain more detailed information since the X-rays sense not only a variation in the scattering factor but also a variation in the lattice spacing[1]. Typical high-angle XRD patterns from TiN/CrN multilayer with TiCrN and CrTiN interlayer with 11nm modulation period are shown in Fig.1. When the out-of-plane lattice spacing of the individual layers are similar, the pattern generally consists of a Bragg peak located at the average lattice spacing of the multilayer surrounded by equally spaced (in reciprocal space) satellite peaks. The lattice mismatch of TiN/CrN is 1.70%. The (111) peak of TiN and CrN with (111)-oriented TiN/CrN multilayer is almost superposition.

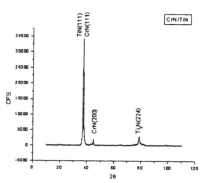

Figure 1: XRD of TiN/CrN multilayer coatings with TiCrN and CrTiN interlayer

Hardness and Adhesion

The nanoindentation hardness curve of TiN/CrN with TiCrN and CrTiN interlayer multilayer are shown in Fig.2. The hardness is various with displacement into surface. The hardness of TiN/CrN multilayer with TiCrN and CrTiN interlayer in Fig.2 are 33GPa at 11nm modulation period and 2μm thickness, which is close to the 36GPa for TiN/CrN coating that has the modulation length of 8nm [15]. For comparison, the measured hardness for TiN and CrN coatings deposited under the similar deposition conditions were 21 and 16 GPa, respectively.

The insufficient adhesion is still one of the most important issues in the development of new coating technologies for hard metal cutting tools. For the multilayer coatings, the adhesion is especially important because it is necessary to the good bonding not only between the coating and the substrate but also between all the layers involved. In the research, scratch tests were employed to investigate the adhesion of the new coatings with interlayer. It was found that under low scratch loads, the sliding of the diamond tip could only induce elastic deformations to the coating-substrate system, leaving no visible surface damages. With an increase in the load force to 60N, cracks appeared on the surface of the coatings. Two types of cracks could be observed: cracks that parallel along the scratches and scratches perpendicular to the scratches transverse with a semicircular morphology (Fig.3a). At higher normal load force up to 80N, only a couple of discontinuous chippings were visible(Fig.3b). This indicates that TiN/CrN multilayered coatings with TiCrN and CrTiN interlayer have a high deformation resistance and good toughness The maximum load of the this coating is 85N, which is much better than the conventional TiN/CrN multilayer coatings.

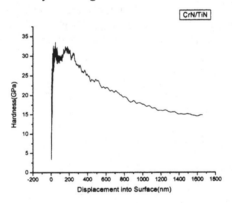

Figure 2: The variation of nanoindentation hardness vs distance to surface TiN/CrN multilayer coatings with TiCrN and CrTiN interlayer.

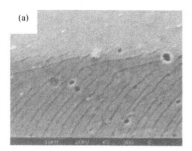
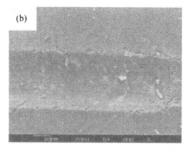

Fig.3 Surface morphologies of TiN/CrN multilayered coatings with TiCrN and CrTiN interlayer after scratching tests showing: (a) cracks and (b) chippings

Effect of Modulation Wavelength on Hardness

Most of the cutting tools contain sharp edges and corners. Therefore, the interaction between the ions and the sharp edge of corners are different than in the case of flat samples, which will affect the coating growth kinetics. Localized changes in the hardness, adhesion, morphology, and composition of coatings on edges and corners have been reported [16–20]. In industry, empirical optimization of the deposition conditions is often done in order to decrease the unwanted effects. In the case of cutting tools coated with TiN/CrN deposited by linear filtered arc ion plating, for example, thickness changes in the appearance of the coatings in the cutting edge region have been observed for negative substrate biases 150 V and temperature 450°C.

The hardness and stress of the edge were more than that of flat surface, and the probability of peeling in edge is also higher. Two methods have been applied to reduce the stress at the edges. One is to adjust on-off ration of pulse bias. As proposed by E. B. Macak et[21,22], in order to decrease the edge-related effects, that negative bias voltage value is down to 75V from 150 in cutting tools coatings TiAlN by magnetron sputtering. However, this method is at the cost of hardness to acquire uniformity of hardness at edge and flat. Another method to reduce the edge effects is to apply compound interlayer between TiN and CrN. In this research, TiN/TiCrN/CrN/CrTiN composite multilayer coatings were formed for trying to broaden modulation period of high hardness of superlattice multilayer coatings and toughness of multilayer system is improved more.

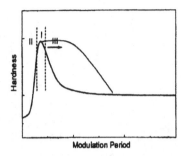

Figure 4: Sketch of extending high hardness modulation period range.

For the most of superlattice multilayer, interface diffusion is inevitable at high temperature. Results of the interdiffusion induced alloying layer are believed to be responsible for the hardness degradation. The composite multilayer is propitious to high working temperature due to composite nitrides TiCrN or CrTiN existing on interface TiN and CrN, which diffusion motivity of metal elements at interface are weakened, in turn, affects the coating properties. The alloying action to nitrides makes multilayer system holding on higher hardness. Further, range of modulation period of the highest hardness is extended to larger single thickness. Fig.4. shows that hardness of superlattice multilayer coating related with modulation period. In the coatings without interlayer, the range of higher hardness modulation period is narrow, generally, from 5nm to 20nm (area I in Fig.4). In the coatings with interlayer, the modulation period of high hardness is extended from I area to III area. Fig.5a shows that hardness of the coating with interlayer keeps about 28GPa when modulation period vary from 11nm to 80nm. At modulation period 200 nm, the coating hardness is still close to 28GPa.

The hardness of the TiN/CrN nano-multilayer coating with TiCrN and CrTiN interlayers increase slinearly with $\Lambda^{-1/2}$, when Λ is more than 100nm. This is consistent with the predictions from Hall-Petch type strengthening mechanism. However, for small modulation periods, a deviation from the linear Hall-Petch relation was observed, as shown in Fig.5b, higher hardness would be expected if the modulation period decreases further.

Hall-Petch relationship relating the grain size of a polycrystalline material to its mechanical properties has been known for many years. The general form of the relationship is given by:

$$H = H_0 + k\,D^{-1/2} \qquad (1)$$

where H is the hardness of polycrystalline material with grain size D, H_0 is the hardness of the same material with large grains, and k is a constant measuring the relative hardening contribution of the grain boundaries. The original model is based on the assumption that dislocations cannot move through grain boundaries, and dislocation pile-ups in one grain must initiate a dislocation source in an adjacent grain. Many authors have used the Hall-Petch relationship to model the hardness behavior of multilayered materials by replacing the grain size with the multilayer period Λ. This may be reasonable if one of the interfaces in each pair of layers is impenetrable to dislocations. However, it should be pointed out that the models on which Eq. (1) derived were based on grain sizes that were large enough to accommodate many dislocations [23]. In a multilayer coating with a bi-layer period of less than 100nm, specially, modulation period less than 20 nm, only a few dislocations are present in a single layer, and therefore, the relationship between hardness and modulation period will deviate from H-P equation

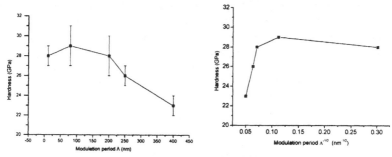

Figure 5: Hardness of multilayered coatings as a function of (a) coating modulation period Λ and (b)$\Lambda^{-1/2}$

Cross-Section Characteristics of TiN/CrN Multilayer Films with TiCrN and CrTiN Interlayer

Fig.6a and Fig.6b show SEM micrographs of fractured and polished cross-sectional views for a TiN/CrN multilayer coatings with TiCrN and CrTiN interlayer. The overall thickness of the coating is 2μm, but the thickness of each ndividual TiN, TiCRN, CrN and CrTiN layers can not be accurately measured from the photographs. By considering the number of layers and total thickness of the coating, the modulation period was determined to be about 80nm (Fig.6a). The morphology of the coating showed no sign of columnar structure, revealing that the grain size of the coating was very small.

After optimizing the linear filtered arc conditions and pulse bias parameter of on-off ratio (duty ratio) of the bias voltage, various magnetron strength, negative bias and duty ratio were used in processes of several depositing stages. Although magnetic filters and substrate pulse bias were applied during the coating process, some small particles still existed in the coatings. As shown in Fig.7, a small particle (<2μm) surrounded by multilayer is observed on the high speed steel substrate. It is discovered that the small particle is embedded in TiN/CrN multilayered coatings with TiCrN and CrTiN interlayer and no interstice is observed between the particle and multilayer film. It is deduced that compact films are formed due to multilayer structure.

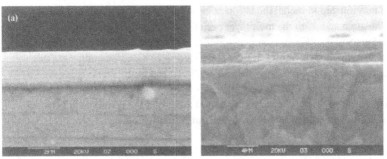

Figure 6: The polished-etched cross-sectional vies (a) and fractured (b) of TiN/CrN multilayered coatings with TiCrN interlayer

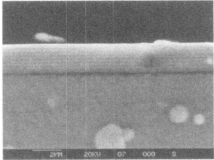

Fig.7 The embedded particle in TiN/CrN multilayered coatings with TiCrN and CrTiN interlayer

Crack propagation tests were conducted by means of three-point bending to investigate the coating toughness. As shown in Fig. 8 that crack deflection is evidently observed on the sample cross-section. It is believed that this is due to the superlattice structure in the multilayer coatings that change direction of crack propagation [24], and then the penetration depth of the cracks is substantially reduced in layered structures. It is also found on Fig. 8 that the micro-pinhole near crack between film and substrate is observed, which is closed due to multilayered structure.

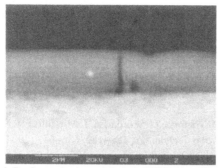

Figure 8: The crack deflection and micro-pinhole in TiN/CrN multilayered coatings with TiCrN and CrTiN interlayer

CONCLUSIONS

In this research, nano-multilayer TiN/CrN coatings with TiCrN and CrTiN interlayers were deposited by linear filtered arc ion plating (LFAIP), on high speed steel M2 substrates. The high hardness modulation period range of this new coating is extended from about 11 nm to 100nm, and the coating hardness is still kept at 25GPa when the modulation period is 200nm. Scratching test shows that the adhesion of this new coating is more than 80 N. Three-point bending test revealed that crack-deflection and pinhole-close effects can significantly improve the toughness of the coating.

REFERENCES

1. P.C. Yashar, and W.D. Sproul, *Vacuum*, **55**,179-190 (1999).
2. U. Helmersson, S. Todorova, S.A. Barnett, J.E. Sundgren, L.C. Markert, and J.E. Greene. *J Appl Phys.*, **62**, 481-484 (1987)
3. M. Nordin, M. Larsson, S. Hogmark, *Surf. Coat. Technol.*, **106**:234-241 (1998).
4. P. Yashar, S.A. Barnett, J. Rechner. W.D. Sproul, *J Vac Sci Technol A*, **16**, 2913-2918 (1998).
5. C. Mendibide, P. Steyer, J.P. Millet, *Surface and Coatings Technology*, **200**, 109-112 (2005)
6. P. Panjan, B. Navinsek, A. Cvelbar, A. Zalar, J. Vlcek, *Surf Coat Technol* **98**, 1497-1502 (1998).
7. P. Panjan, B. Navinsek, A. Cvelbar, A. Zalar, I. Milosev., *Thin Solid Films*, **281-282**, 298-301 (1996).
8. Nordin, M., M. Herranen, and S. Hogmark. Thin Solid Films, 1999. 348(1-2):202
9. Q. Yang, D.Y. Sep, and L.R. Zhao, *Surface and Coatings Technology*. **177-178**, 204-208 (2004).

10. C. Mendibide, P. Steyer, C. Esnoouf, P. Goudeau, D. Thiaudiere, M. Gailhanou and J. Fontaine, *Surface and coatings technology*, **200**, 165-169 (2005)
11. W.D. Munz, D. B. Lewis, P. Eh. Hovsepian, C. Schonjahn, A. Ehiasarian, and I. J. Smith, *Surface Engineering*, **17**, 15-27 (2001).
12. E. B. Macak, J. M. Rodenburg, *J. Vac. Sci. Technol.A*, 22, 1195-1199 (2004).
13. E. B. Macak, W.D. Munz, and J. M. Rodenburg, *Surface Engineering*, **19**, 310-314. (2003)
14. K.D. Bouzakisa, S. Hadjiyiannisa, G. Skordarisa, I. Mirisidisa,N. Michailidisa, G. Erkens, *Surface & Coatings Technology*, **188-189**, 636-643 (2004).
15. Harish C. Barshiliaa, Anjana Jainb, K.S. Rajama, *Vacuum*, **72**, 241-248 (2004).
16. B. O. Johansson, J.-E. Sundgren, H. T. Hentzell, and S.E. Karlsson, *Thin Solid Films*, **111**, 313-322 (1984).
17. S. S. Kim, J. G. Hang, and S. Y. Lee, Deposition behaviours of CrN films on the edge area by cathodic arc plasma deposition process, *Thin Solid Films*, **334**, 133-139 (1998).
18. R. Novak, I. Kvasnicka, D. Novakova, and Z. Mala, *Surf. Coat. Technol.*, **114**, 65-69 (1999)
19. H. A. Jehn, B. Rother, H. Kappl, and G. Ebersbach, *Surf. Coat. Technol.*, **94**, 232-236 (1997).
20. E. B. Macak, W.D. Munz, and J. M. Rodenburg, *Surf. Coat. Technol.* **151**, 349-354 (2002).
21. E. B. Macak, W.D. Munz, and J. M. Rodenburg, *J. Appl. Physics*, **94**, 2829-2836 (2003).
22. E. B. Macak, W.D. Munz, and J. M. Rodenburg, *J. Appl. Physics*, **94**, 2837-2844 (2003).
23. G.E. Dieter, *Mechanical Metallurgy*. McGraw-Hill, Inc., (1986).
24. H.Hlleck and V.Schier, *Surface Coatings and Technology*, **76-77**, 328-336 (1995).

DEVELOPMENT OF A DUPLEX COATING PROCEDURE (HVOF AND PVD) ON TI-6AL-4V SUBSTRATE FOR AUTOMOTIVE APPLICATIONS.

E. Bemporad, M. Sebastiani[*], F. Carassiti
University of Rome "ROMA TRE", Mechanical and Industrial Engineering Dept
Via della Vasca Navale 79
Rome, 00146, Italy

F. Casadei, R. Valle
Centro Sviluppo Materiali SpA, Surface Engineering and Ceramic Unit
Via di Castel Romano 100
Rome, 00128, Italy

ABSTRACT

Titanium and its alloys are extensively used in aerospace and mechanical applications because of their high specific strength and fracture toughness. On the other hand, titanium alloys often show poor resistance to sliding wear, low hardness and load bearing capacity, so that surface performance improvement is often recommended.

Present work deals with design, production and characterization of a duplex coating system for Ti6Al4V components (i.e. crankshafts, piston rings, connecting rods) consisting of an high velocity Oxygen fuel (HVOF) WC-Co interlayer and a Ti/TiN multilayer (two layer pairs) deposited by cathodic arc evaporation - physical vapour deposition (CAE-PVD).

A preliminary coating design was carried out, based on Finite Element simulation of residual stress fields on the PVD coating for a range of configurations of its multilayered structure (Ti buffer layer position and thickness). After the choice of the optimal PVD coating configuration, the influence of HVOF WC-Co substrate roughness on PVD adhesion on cylindrical components was analysed, with the aim of optimising the cost to performance ratio of the coating system.

Morphological properties of the produced coatings (thickness, roughness, grain size, interfaces) were measured by means of Scanning Electron (SEM) and Focused Ion Beam (FIB) microscopy techniques, while mechanical properties were investigated using Rockwell C adhesion test, micro-scratch, micro- and nano-indentation techniques.

The use of a multilayer Ti/TiN PVD coating, which thicknesses and Ti distribution were suggested by Finite Element Modelling, lead to a significant increase (about 60%) in adhesion to the HVOF coating, compared to monolayer TiN, without reduction in superficial hardness.

Furthermore, the maximum allowed HVOF coating surface roughness (R_a) for an optimal adhesion of the PVD coating has been evaluated.

INTRODUCTION

An excellent combination of high specific strength, fracture toughness, corrosion resistance and thermal stability makes Titanium and its alloys particularly suitable candidates for biomedical, aerospace and extreme mechanical applications.

Titanium alloys do however exhibit some disadvantages: low resistance to sliding wear; low hardness; low load bearing capacity; and poor adhesion with respect to coatings.

---

[*] Corresponding author: marco.sebastiani@stm.uniroma3.it

Plasma nitriding and PVD coating are commercial methods for improving properties in those applications which involve high contact stresses and severe sliding wear[1-2].

However, a very thin, hard layer on a Titanium alloy substrate (even if hardened by plasma nitriding) cannot lead to a mechanically improved structure in terms of load bearing capacity: differences in both coating and substrate hardness and stiffness do not provide a good distribution of contact stresses[1], while the presence of localised cracks under contact or tribological loads can generate galvanic corrosion between substrate and the PVD coating[3-4].

An optimised duplex coating may contain a harder and also stiffer (compared to the nitrided layer) interlayer, which provides a better distribution of contact stresses, avoiding plastic deformation of the substrate and brittle failure of the coating. Bemporad et. Al.[5] showed in a recent work that an effective configuration in terms of enhanced load bearing capacity on Ti6Al4V substrates can be achieved by a duplex coating system, consisting of an HVOF thermally sprayed WC-Co thick interlayer and a cathodic arc evaporation (CAE) PVD (TiN or CrN) thin coating.

Adhesion can be enhanced by deposition of multilayered PVD metal/ceramic systems[6-9]: as an example, several studies[6-8] showed that the interposition of a Ti interlayer with a thickness of 0.5–1.5 µm accommodates the PVD TiN internal stresses and allows thicker composite coatings to be produced, with significant improvements in toughness, adhesion, impact resistance and corrosion resistance; however, the presence of a relatively thick Ti buffer layer usually involves a significant reduction in hardness and wear resistance of the PVD coating, especially in the case of a hard PVD coating on a soft Ti-alloy substrate.

It is therefore clear that a deeper optimisation of coating procedures for Ti alloy based components is required, with respect to their surface mechanical performances and corrosion resistance.

The idea for the present research activity was therefore to develop and optimise a coating system for Ti6Al4V components, consisting of a Ti/TiN (two layer pairs) CAE-PVD coating on a HVOF WC-Co thick interlayer, with the aim of increasing adhesion and load bearing capacity of the PVD coating.

All experimental work consisted of three different steps: (1) design optimisation procedure of the PVD coating (parametric study of the influence of the Ti buffer layer position and thickness on interfacial stress peak, based on finite element simulation of residual stress); (2) produced coatings morphological, microstructural and mechanical characterisation, (3) study of the influence of the HVOF surface roughness on PVD coating adhesion and final coating procedure application on crankshafts for high performance engines.

COATING DESIGN AND OPTIMISATION

Starting from results of a previous experimental work by the same authors[5], coating optimisation consisted of a Finite Element based design of the Ti/TiN (two layer pairs) PVD coating, based on residual stress evaluation for several configurations of its multilayered structure.

The objective of this numerical study was to determine the influence of the Ti buffer layer position and thickness on the interfacial residual stress field due to deposition processes.

Delamination failure of PVD coatings is generally caused by high interfacial residual stress fields resulting from the deposition process parameters and from poor adhesion.

Residual stresses in PVD coatings[10-14] arise from the contribution and interaction of two main factors: (A) thermal stress[14] (σ_{th}), arising from differences in thermal expansion coefficient between coating and substrate during final cooling from deposition to room temperature, and (B) intrinsic stress[10] (σ_i), which occur as a consequence of deposition parameters induced crystallographic growth orientation; according to models available in literature, they can be expressed as a function of the ionic and atomic flux arriving at the substrate during deposition, and of the energy distribution of the bombarding ionic species[10].

Usually the resulting residual stress field is approximately calculated as follows:

$$\sigma_{tot} = \sigma_i + \sigma_{th} \tag{1}$$

CAE-PVD coatings always show high in-plane compressive intrinsic stress[11-13], which necessarily involves high normal to surface tensile stress.

The use of ductile Titanium interlayers in multilayer Ti/TiN coatings can be effective for the residual stresses relaxation, in order to improve coating-substrate interfacial adhesion and impact wear resistance, but the presence of a Ti interlayer involves also a reduction in mechanical properties of the coating[9], such as hardness and sliding and abrasive wear resistance.

In the case of HVOF coatings, residual stresses also arise from two sources[14-16]: the quenching stress (σ_q), due to the instantaneous cooling of impacting droplets, and thermal stress, due to differential thermal contraction.

Quenching stresses can be evaluated by the following equation[14-16]:

$$\sigma_q = E_C^* \int_{T_s}^{\beta T_m} \alpha_C (T) dT \tag{2}$$

where T_m is the melting temperature of the sprayed material, T_s is the deposition temperature, E_C^* is the actual elastic modulus of the coating, as determined by experimental measurements, and β a temperature reduction coefficient (approximately equal to 0.6) adopted to take into account stress relaxation phenomena such as yielding or creep[16]. In case of ceramic-based coatings, a marked reduction in quenching stress has been observed[16], due to extensive micro-cracking of individual lamellae after droplet impact. As an example, a residual tensile stress of about 10 MPa can be experimentally evaluated by in situ curvature measurements for plasma-sprayed alumina[16], while quenching stresses of the order of GPa would be obtained by applying equation (2), that is considered to be valid only for metallic sprayed materials.

Considering a duplex HVOF - PVD coating system, complex residual stress interactions are likely to coexist; by adopting realistic simplification hypotheses, described hereafter, a predictive model can be developed for qualitative stress estimation and process optimisation.

Using the commercial Finite Element software, ABAQUS6.5, the duplex systems under examination were modelled as explained in figure 1: all models comprise a plane disk (quadratic, coupled temperature-displacement axisymmetric elements, figure 1) in which the interfaces between coatings are assumed to be perfect, materials mechanical and thermal properties (table I) are assumed to be temperature dependent. Composite material properties were adopted for the HVOF WC-Co coating. Elastic properties of both layers were experimentally determined as reported in the following chapter; plastic behaviour of the substrate is also taken into account. Adopted material properties are summarised in table I.

Several different systems were modelled, in order to investigate the influence of the Ti buffer layer through-thickness position on residual stresses (see table I and figure 1), as follow:

- WC-Co HVOF coating thickness was fixed at 400 μm;
- Overall thickness of the PVD coating was fixed at 6 μm;
- All PVD Ti/TiN coatings modelled consisted of two layer pairs (see figure 1(a));
- Position of the Ti buffer layer was varied from 0,5 to 5,5 μm, assuming as zero the PVD/HVOF interface, with intervals of 0,25 μm;
- Thickness of the Ti buffer layer was varied between 100 and 300 nm;
- A model for the same thickness monolayer TiN coating have been realised as well;

Thirty-one different FE models were therefore realised, all following the same algorithm reported in figure 1(b).

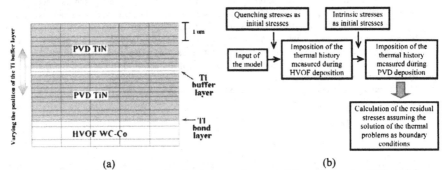

(a)   (b)

Figure1. (a) Two-dimensional axisymmetric coupled thermal-structural FE model realised with varying position and thickness of the Ti buffer layer; (b) adopted algorithm for all simulations

The model simulates the thermal history due to both HVOF and PVD deposition: the code imposes the final cooling from deposition temperatures of both deposition stages and calculates the resulting residual stress field (see algorithm in figure 1(b)).

The intrinsic stress for PVD and the quenching stress for HVOF were assumed as initial stresses (using an existing ABAQUS subroutine): a value of -3 GPa (typical for CAE-PVD coatings[10, 12-13]) was imposed as the intrinsic residual stress for the PVD coating, whereas a value of +70 MPa[14] was adopted for the quenching stress of the WC-Co HVOF coating.

In-plane, normal to the surface and shear stresses were calculated and analysed for each model, as a function of the Titanium interlayer thickness and position.

Table I. Material properties (T = 293K) adopted for FE modelling

	E [GPa] - ν	Yield Strength [MPa]	CTE [10^{-6} C^{-1}]	Mechanical Behaviour
TiN	545* - 0,25	-	9,4	Perfectly Elastic
WC-17%Co	280/320** – 0,2	1500	5,45	Plastic Hardening
Ti6Al4V[24]	114 – 0,35	880	8,6	Plastic Hardening

*measured by nano-indentation test (Oliver & Pharr method)
**normal/in-plane values, measured by Knoop micro-indentation (500 gf, Marshall model[20])

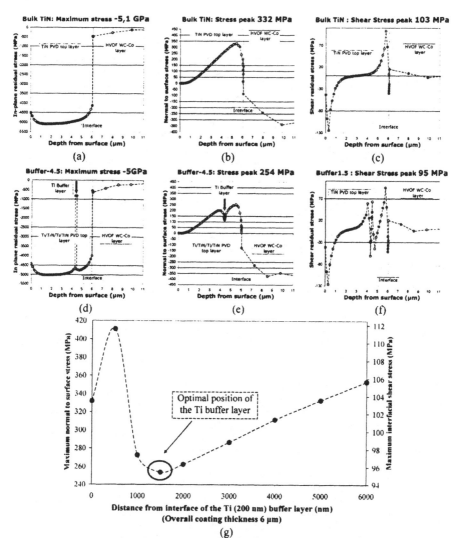

Figure 2 (a) In-plane, (b) normal to surface and (c) shear predicted residual stresses for model Bulk (monolayer TiN-PVD 6μm); (d) In-plane, (e) normal to surface and (f) shear predicted residual stresses for model with 200 nm buffer layer placed at 1,5 μm from HVOF/PVD interface; (g) Maximum normal to surface (left axis) and shear (right axis) stress peaks as a function of the Ti buffer layer position (from PVD/HVOF interface).

EXPERIMENTAL DETAILS

Coatings deposition

Starting from results of simulations, described hereafter, duplex systems were obtained by the sequential deposition of a WC-17%Co thick coating, by a High Velocity Oxygen Fuel technique followed by a multilayer Ti/TiN CAE-PVD coating (samples geometries: plane disk and cylinder, 20 mm and 15 mm diameter respectively.

Deposition of HVOF coating was performed using a commercial liquid fuel JP-5000 Hobart-Tafa HVOF apparatus (process parameters: working distance: 380mm; Oxygen flow-rate $16 \cdot 10^{-2}$ m^3/s; kerosene flow-rate $7,57 \cdot 10^{-6}$ m^3/s, Work temperature 400 °C); average thickness of WC-Co coating was of 400 μm, measured by cross-section optical microscope analysis.

Microstructure of the HVOF coating, porosity and phases distribution were evaluated by image analysis on polished samples[5].

In order to prepare the HVOF coating surface for the subsequent deposition of the Ti/TiN coating by PVD, an accurate lab-scale polishing sequence was set up, following procedures described in previous work by the authors[5].

Several substrates of cylindrical geometry were also adopted, as reported in table III: in this case, an industrial polishing procedure was adopted, using a commercial cylinder grinder; in order to find out the influence of the HVOF roughness on the PVD coating adhesion, cylindrical samples have been polished at three different levels of surface roughness (see table III).

PVD monolayer and multilayer coating deposition was performed using the process parameters summarised in table II; note that the same process parameters were adopted for both types of coating deposition.

Table II. process parameters for PVD deposition

	Pressure [Pa]	Deposition Temperature [°C]	Bias [V]	Current [A]
TiN layers	1,5	450	135	50
Ti buffer layer	0,8	450	135	50

Coatings morphology and microstructure

Morphological characterization of coatings (including coating thickness, microstructure and crystallite size, surface and layer interface analyses) were performed using digital optical microscopy, Scanning Electron Microscopy (SEM) and Focused Ion Beam (FIB) techniques.

By the use of ion milling techniques artefact-free cross sections and TEM lamellae were obtained and a detailed analysis of coating microstructure and interfaces was performed.

Coatings mechanical properties

Adhesion of the PVD coating was quantitatively evaluated (critical loads: L_{c1} cohesive failure; L_{c2} adhesive failure; L_{c3} delamination) by means of scratch testing[17]: the following test parameters were adopted: progressive loading scratch test (PLST), loading rate 10 N/min, table traverse speed 10 mm/min, according to the standard UNI EN 1071-3.

Adhesion and load carrying capacity of the coating PVD coating were also qualitatively evaluated using standard Rockwell C indentation test: this procedure (standard DIN CEN/TS 1071-8)[18], based on image analysis of fracture behaviour of the coating under a high load indentation test, allows to obtain an effective comparison between different PVD coatings on the HVOF coating.

Intrinsic hardness and reduced modulus of the PVD coating have been measured by means of nano-indentation testing: 50 indentation tests have been performed adopting the following test parameters: 30 mN maximum load, 1,33 mN/s loading rate, 10 s hold at peak load for creep, 20 s hold at 90% for thermal drift correction. Oliver & Pharr method has been adopted for hardness and elastic modulus evaluation.

WC-Co coating elastic modulus and anisotropy was evaluated by means of Knoop micro-indentation tests, adopting the model proposed by Marshall[19], which correlates the hardness to modulus ratio with the elastic recovery of the in-surface dimensions of a Knoop indentation; following a procedure already reported in literature[20] it is also possible to evaluate coating mechanical anisotropy by performing Knoop tests (500 gf applied load) along several selected directions: in this case the in-plane and normal to surface modulus have been evaluated.

Coatings composite hardness was measured by the use of a standard Vickers micro-hardness tester (applied loads 0,5-300 gf). Low loads (<10 gf) indents shape evaluation was carried out by means of atomic force microscopy. The obtained value for each load corresponds to the mean of six measures. In order to extrapolate film intrinsic hardness, the Jonsson & Hogmark and the Chicot & Lesage models have been used[21-22]; since the HVOF coating is much thicker than the PVD ones, the influence of the Ti-6Al-4V substrate was assumed to be negligible for the in plane micro-hardness tests.

RESULTS

Finite Element Modelling based design of the Ti/TiN coating.

Results for Finite Element Simulations of residual stress distribution are shown in terms of in-plane (radial), normal to surface (axial) and shear residual stress components profile through the coating thickness and with respect to the Ti buffer layer position.

In particular, figure 2(a)-2(c) show the residual stresses predicted distribution in the case of a bulk 6μm monolayer of TiN (WC-Co substrate): results show a high compressive in-plane stress (node path in correspondence of the disk axis), resulting from both intrinsic and thermal stresses, and a correlated tensile normal to surface stress (node path in correspondence of the sample edge), which is obviously zero at the free surface and show a peak in correspondence of the PVD/HVOF interface.

Figure 2(d)-(f) show the in-plane, normal to surface and shear predicted stress component variation for a two layer pairs Ti/TiN with buffer layer (200 nm thick) placed at 1,5 μm from the HVOF/PVD interface: in this case is evident the reduction in normal to surface stress peak (about 24 % compared to same thickness monolayer TiN); a similar reduction is also predicted for the interfacial shear stress peak (figures 2(c)-2(f)).

Analogous stress fields have been obtained for all other models realised, being different the interfacial stress peak value, which is reported in figure 2(g) as a function of the Ti buffer layer position.

It can be seen that a minimum exists when the buffer layer is placed at 1,5 μm from HVOF/PVD interface (overall PVD coating thickness 6μm, buffer layer thickness 200 nm): this configuration (whose predicted stresses are reported in figures 2(d)-2(f) is therefore expected to guarantee the best stress reduction and the optimal adhesion of the PVD coating.

Coatings deposition
Starting from results of simulations produced coatings are then as follows (see table III):

- Duplex monolayer 4 µm TiN on 400 µm polished HVOF WC-Co on planar Ti6Al4V substrate (sample code *PBRa1*);
- Duplex multilayer PVD Ti/TiN (two layer pairs Ti/TiN with buffer layer (200 nm thick) placed at 1,5 µm from the HVOF/PVD interface) on 400 µm polished WC-Co HVOF on planar Ti6Al4V substrate (sample code *PMRa1*);
- Three duplex systems (identical to the previous one) on cylindrical substrate, with different HVOF coating surface roughness (samples codes *CMRa1, CMRa2, CMRa3*, as reported in table III).

Table III Summary of all produced coatings

Sample code	Substrate geometry	HVOF thickness [µm]	HVOF surface roughness [µm]	PVD coating thickness [µm]	Ti bond-layer thickness [nm]	Ti buffer-layer thickness [nm] and position [µm]
PBRa1	Planar	400	0,013	4	50	Bulk TiN
PMRa1	Planar	400	0,013	6	50	200 ; 1,5
CMRa1	Cylindrical	400	0,016	6	50	200 ; 1,5
CMRa2	Cylindrical	400	0,052	6	50	200 ; 1,5
CMRa3	Cylindrical	400	0,068	6	50	200 ; 1,5

Coatings morphology and microstructure
The microstructure of the HVOF WC-Co coating is shown in figure 4(a): the measured mean porosity was found to be 1,7%; further details are also reported in [5]. Roughness of the HVOF coating, measured by stylus profilometer for all realised samples is reported in table III.

Liquid Nitrogen fractured surface of the produced duplex system is reported in figure 4(b), where a marked columnar microstructure is evident for the TiN layers, and no delamination cracks were detected.

Focused ion Beam microstructural analysis is reported in figures 5(a)-5(d): figure 5(a) show a typical cross section obtained by site-specific ion milling for the cylindrical sample *CMRa1*, while figure 5(b) show the PVD coating microstructure: columnar growth orientation, growth and grain size of the PVD coating are visible; figure 5(c)-5(d), obtained by in-situ STEM observation after FIB thinning, show a detail of the buffer layer and of the HVOF/PVD interface.

Table IV. Results summary for the mechanical characterization of the PVD coatings.

Sample code	Intrinsic Hardness (J-H model) Meyer HV_0 ; n [GPa] ; #	Intrinsic Hardness (C-L model) Meyer HV_0 ; n [GPa] ; #	Reference Intrinsic Hardness* [GPa]	HRC (1470N) adhesion test Class #	Critical load adhesive failure L_{c3} (Scratch test) [N]	TiN coating elastic modulus (nanoindentation) [GPa]
PMRa1	27,273 ; 1,972	26,452 ; 1,990	25,5	1	29,0	545
PBRa1	35,075 ; 1,990	32,387 ; 2,010	30,9	1	18,2	-
CMRa1	-	-	-	1	29,0	-
CMRa2	-	-	-	1	28,5	-
CMRa3	-	-	-	2	23,0	-

*Calculated at indentation depth h equal to one tenth of the coating thickness t: $h=t/10$ (Jonsson model).

(a)　　　　　　　　　　　　　　　　(b)

Figure 4. (a) Surface plan-view SEM observation (20 kV; SE; 3000 X) of the HVOF WC-Co interlayer and (b) liquid Nitrogen fractured surface cross-section SEM observation (20 kV; SE, 7000 X) of sample *PMRa1*.

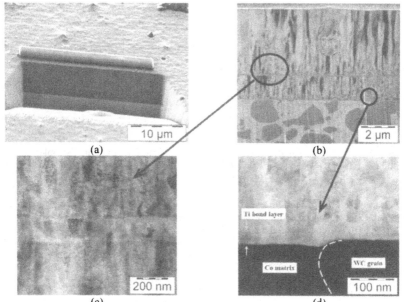

(a)　　　　　　　　　　　　　　　　(b)

(c)　　　　　　　　　　　　　　　　(d)

Figure 5. (a) sample *CMRa1* FIB cross section (in-situ SE-SEM: 5 kV; 3,700 X), (b) PVD coating morphology and microstructure (FIB CDEM: 30kV; 24,000X), (c) detail of the Ti-buffer layer (in-situ STEM, after FIB thinning, 30 kV; 250,000 X) and (d) detail of the PVD/HVOF interface (STEM: 30 kV; 700,000 X).

Coatings mechanical properties

Results of the mechanical characterisation are summarised in table IV.

Scratch results (also reported in figure 6) show a significant increase (about 60%) of critical loads for the Ti/TiN multilayers, compared to the monolayer TiN: it also noteworthy that the adhesive failure of the multilayered optimised Ti/TiN coating (figure 6(a)) is much less evident than the ones observed for Bulk TiN (figure 6(b)), which is complete delamination; furthermore, in case of cylindrical samples with different HVOF roughness, it can be observed a significant reduction of critical load for the higher roughness value considered (sample CMRa3, Ra=0,07 μm); conversely, no changes have been observed until Ra=0.05 μm (CMRa1, CMRa2).

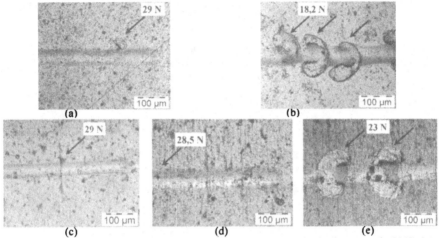

Figure 6. Failure modes of analyzed coatings under scratch testing: (a) planar sample PMRa1 (multilayer Ti/TiN), (b) planar sample PBRa1 ("Bulk" TiN), (c) sample CMRa1, (d) sample CMRa2 (e) sample CMRa3.

Qualitative HRC adhesion test (figure (7)) resulted to be in accordance with previous results[5], duplex coatings showed a similar behaviour as the bulk TiN (according to the standard DIN CEN/TS 1071-8, class 1) on the same substrate.

Vickers Hardness measurement and modelling lead to the results reported in table VI for samples PMRa1-PBRa1 (planar substrate), according to the Meyer presentation of the Indentation Size Effect (ISE)[23]:

$$HV = HV_0 \cdot d^{n_m}$$ (3)

Where d is the indent dimension; HV_0 represents the hardness for infinitesimal applied loads, while n_{HT} (Meyer Index) can be related to the material hardening behaviour (being $n_{HT} = 2$ in case of perfectly plastic mechanical behaviour).

It can be observed that intrinsic hardness of the multilayered Ti/TiN coating (considered as a bulk in the J-H and C-L models) is lower than the one obtained for monolayer TiN.

Elastic properties evaluation results for TiN and WC-Co coating, performed by nano-indentation and Knoop micro-indentation tests are reported in table 1.

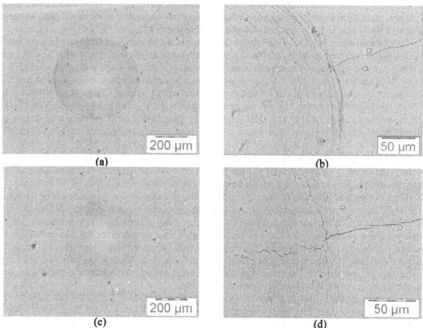

Figure 7. Toughness and Load Bearing capacity evaluation by HRC (1471 N) adhesion test: (a,b) sample *PBRal*, (c,d) sample *PMRal*.

DISCUSSION

The goal of the present work was to investigate the feasibility of a new duplex coating procedure, based on a PVD Ti/TiN multilayer top-coat on a thick WC-17%Co interlayer for Ti-6Al-4V components.

Figure 2 reports the predicted in-plane (radial), normal (axial) and shear stress distributions for the Bulk and the optimised Ti/TiN PVD coatings: in-plane stress (about -5GPa) resulted to be in accordance with results available in literature (XRD-$\sin^2\psi$ method) for similar coatings[12-13]. On the other hand, predicted normal to surface stress is obviously zero at the free surface, but increases towards HVOF/PVD interface, showing a peak in its correspondence; however, in the case of multilayer systems both this stress peak and the interfacial stress gradient are reduced, compared with the monolayer system (figure 2(b)-(e)): this can be explained in terms of plastic strain of the ductile Ti buffer layer and consequent stress rearrangement.

It is also evident that the value of this interfacial stress peak changes with the position of the Ti buffer layer (figure 2(g)) showing a minimum when the buffer-layer is placed at 1,5 μm from interface. This result can be explained considering the predicted through-thickness stress variation: if the stress increases towards the (HVOF) interface, it is likely that a lower position of Ti interlayer will be more effective in terms of stress relaxation; on the other hand, if the buffer layer was placed too much in proximity to the interface, an increasing in the stress peak would be observed, likely due to high plastic strain of the two adjacent Ti layers.

The maximum normal stress (figure 2(b)-2(e)) is not exactly placed at interface: at first it was supposed that this could be due to an inadequate mesh refinement, but after a further series of simulations it was found that adopted mesh size was sufficient to avoid influence on results.

FIB analyses, reported in figure 5, gave lot of information about coating microstructure and its growth mechanisms: it can be observed (figure 5(b)) a marked columnar structure for the TiN layers, indicating coating texture; figure 5(c) report a detail (obtained by in-situ STEM, after FIB lamella thinning) of the Ti buffer layer: it is clear that this Ti interlayer promotes re-nucleation of the TiN phase: this could be one of the main reasons of the improved film adhesion.

Figure 5(d) shows a detail of the HVOF/PVD interface: a different coating growth mechanism seems to exist whether PVD grain grow on a WC or Co grain: TEM-SAED analyses are currently ongoing in order to find out the actual composition of the HVOF/PVD interface.

As expected, adhesion of the PVD coating is enhanced by the presence of the Titanium interlayer: micro-scratch testing confirmed that the optimised duplex system adopted (sample *PMRa1*) involves a remarkable increase in adhesion; the critical load L_{C3} measured for the duplex-multilayer coating was found to be 60% higher than obtained for the duplex-monolayer coating (*PBRa1*). It is also evident from figure 5 that different failure mechanisms occur for buffered systems: in case of bulk TiN a stronger interfacial delamination has been observed than in the case of the multilayered coatings: this is likely due to stress relaxation and increased deformability given by the Ti buffer layer interposition.

A study of the influence of the HVOF surface roughness on PVD/HVOF adhesion on real cylindrical components has been carried out: it is well known from literature[25] that a mirror-like surface polishing is required for an optimal adhesion of PVD coatings, being growth mechanisms and direction of columnar grains strictly correlated to the substrate geometry at micrometric scale and to the consequent free surface energy distribution: results for cylindrical samples showed no significant reduction in critical loads (see table IV and figure 6) for HVOF roughness lower than 0,06 μm: this mean that production costs can be significantly reduced by the adoption of WC-Co coating polished to 0,05 μm Ra; adhesion of multilayered coating is anyhow always higher than monolayer TiN.

Furthermore, the duplex coatings approach generally provided an equivalent load bearing capacity because of the high hardness and stiffness of the WC-Co thick interlayer[5]; in the case of the buffered PVD coating, load bearing capacity is even improved after the introduction of the Ti interlayer: as shown in figure 6(b)-(d) radial cracks are less pronounced and shorter in the case of the multilayer PVD coating (sample *PMRa1*) this could be due to the optimisation procedure which allowed to introduce a minimum quantity of Titanium, with an increase in PVD coating deformability without significant reductions in its stiffness and hardness (see table IV, hardness results); other experimental studies[9] adopted much thicker buffer layer (more than 500 nm) placed in middle position: in this case a stronger reduction in coating hardness and wear resistance would be observed.

CONCLUSIONS

A duplex coating system for Ti6Al4V substrates, comprising a 6 μm PVD Ti/TiN multilayer (two layer pairs) on an 400 μm HVOF WC-17%Co thick interlayer has been designed, produced and characterised.

A design procedure, based on Finite Element Modelling of residual stresses arising from deposition processes, allowed to find the optimal configuration (i.e. Ti buffer position and thickness) of the PVD coating.

Scratch test adhesion measurements confirmed simulations showing a significantly increased adhesion of the PVD coating (60% increased critical load compared to monolayer).

The introduction of such a thin Ti buffer layer (200 nm) did not involve an excessive reduction in hardness, whilst improved toughness and load bearing capacity were observed.

A study of the influence of HVOF WC-Co roughness on adhesion in the case of both planar cylindrical shaped samples allowed to find the critical roughness value (Ra 0,05 μm) for production cost to performance ratio optimisation.

The whole coating cost will be anyway higher than for a traditional duplex process (nitriding plus PVD) and also applicable for only a limited range of component geometry, but the optimisation procedure performed, united with the increase in life-time and overall performances (in terms of load bearing capacity and durability) make such layered systems suitable candidate for light-alloy components coating in the race automotive field.

REFERENCES

[1] T. Bell, H. Dong and Y. Sun, "Realising the potential of duplex coatings", *Tribology International*, **31**, 127-37 (1998)

[2] G.W. Critchlow, D.M. Brewis, "Review of surface pretreatments for titanium alloys", *Int.J. Adhesion and Adhesives*, **15** 161-72 (1996)

[3] J. Komotori, B.J. Lee, H. Dong, P.A. Dearnley, "Corrosion response of engineered titanium alloys damaged by prior abrasion", *Wear*, **251**, 1239–49 (2001)

[4] C. Liu, Q. Bi, A. Matthews, "Tribological and electrochemical performance of PVD TiN coatings on the femoral head of Ti-6Al-4V artificial hip joints", *Surf. Coat. Technol.*, **163 –164** 597-604 (2003)

[5] E. Bemporad, M. Sebastiani, D. De Felicis, F. Carassiti, R. Valle, F. Casadei, "Production and characterisation of duplex coatings (HVOF end PVD) on Ti-6Al-4V substrate", *Thin Solid Films*, **515**, 186-194 (2006).

[6] H. Holleck and H. Schulz, "Preparation and behaviour of wear-resistant TiC/TiB$_2$, TiN/TiB$_2$ and TiC/TiN coatings with high amounts of phase boundaries", *Surf. Coat. Technol.*, **36**, 707-14 (1988)

[7] A. Leyland, A. Matthews, "Hybrid techniques in surface engineering", *Surf. Coat. Technol.* **71**, 88-92 (1995)

[8] G.S Kim; S.Y Lee, J.H. Hahn, B.Y. Lee, J.G. Han, J.H. Lee, S.Y. Lee, "Effects of the thickness of Ti buffer layer on the mechanical properties of TiN coatings", *Surf. Coat. Technol.*, **171**, 83–90 (2003).

[9] Y.L. Su, W.H. Kao, "Optimum multilayer TiN-TiCN coatings for wear resistance and actual application", *Wear* **223**, 119–30 (1998)

[10] Y. Pauleau, "Generation and evolution of residual stresses in physical vapour-deposited thin films", *Vacuum*, **61**, 175-81 (2001).

[11] V. Teixeira, "Mechanical integrity in PVD coatings due to the presence of residual stresses", *Thin Solid Films* **392**, 276-81 (2001).

[12] F.R. Lamastra , F. Leonardi, R. Montanari, F. Casadei, T. Valente, G. Gusmano, "X-ray residual stress analysis on CrN/Cr/CrN multilayer PVD coatings deposited on different steel substrates", *Surf. Coat. Technol.* **200**, 6172–6175 (2006).

[13] M. Gelfi, G.M. La Vecchia, N. Lecis, S. Troglio, "Relationship between through-thickness residual stress of CrN-PVD coatings and fatigue nucleation sites", *Surf. Coat. Technol.* **192**,263–268 (2005).
J. Stokes, L. Looney, Residual stress in HVOF thermally sprayed thick deposits, *Surf. Coat. Technol.*, **177 –178**, 18 (2004).

[14] S. Kuroda, T.W. Clyne, "The quenching stress in thermally sprayed coatings", *Thin Solid Films*, **200 (1)**, 49-66 (1991).

[15] T.W. Clyne, S.C.Gill, "Residual Stresses in Thermal Spray Coatings and Their Effect on Interfacial Adhesion: A Review of Recent Work", *Journal of Thermal Spray Technology*, **5 (4)** 401-18 (1996).

[16] European standard UNI EN 1071-3.

[17] German standard DIN CEN/TS 1071-8.

[18] D. B. Marshall, T.Noma, A. G. Evans, "A simple method for determining elastic-modulus-to-hardness ratios using Knoop indentation measurements", *J. Am. Cer. Soc.*, Vol. **65** (10) C175-C176 (1982).

[19] S-H. Leigh, C-K. Lin, C.C. Berndt, "Elastic response of thermal spray deposits under indentation tests", J. Am. Cer. Soc., Vol. 80 (8) 2093-192 (1997)

[20] B. Jönsson, S. Hogmark, Hardness measurements of thin films", *Thin Solid Films* **114**, 257–69 (1984).

[21] D. Chicot, J. Lesage, "Absolute hardness of films and coatings", *Thin Solid Films*, **245**, 123-30 (1995).

[22] D. Tabor, *Hardness of Metals* (Clarendon Press, Oxford, 1951).

[23] CES. The Cambridge Engineering Selector, http://www.grantadesign.com.

[24] B. Casas, M. Anglada, V. K. Sarin, TiN coating on an electrical discharge machined WC-Co hardmetal: surface integrity effects on indentation adhesion response, *J Mater Sci*, **41** 5213-19 (2006)

NOVEL COATINGS OF CEMENTED CARBIDES BY AN IMPROVED HVOF SPRAYING PROCESS

Makoto Watanabe, Pornthep Chivavibul, Jin Kawakita, Seiji Kuroda
Composites and Coatings Center, National Institute for Materials Science
Sengen 1-2-1
Tsukuba, Ibaraki, Japan 305-0047

ABSTRACT

High velocity oxy-fuel (HVOF) spraying has been widely used to deposit coatings of cemented carbides such as WC-Co for various applications. When compared with sintered bodies of WC-Co, however, HVOF sprayed WC-Co coatings can be harder but are much inferior in terms of fracture toughness. This is due to the formation of brittle phases such as W_2C and the dissolution of WC into the Co binder phase due to the high flame temperature of HVOF over 2000K. We have recently developed an improved HVOF spray process called the two-stage HVOF (Warm Spray), by which a cold gas can be introduced to control the gas temperature below the melting point of the binder metal phase while maintaining the very high gas velocity. By applying this process to WC-12Co powder, it was demonstrated that essentially degradation-free microstructure could be retained in the sprayed coatings. Preliminary results of the measurements of the mechanical properties of such coatings are reported in this paper.

INTRODUCTION

WC-Co coatings have been used to enhance the wear resistance of various engineering components due to their high hardness and moderate toughness compared with other coating materials. So far high velocity oxy-fuel (HVOF) spraying is known as the most suitable technique to fabricate WC-Co coatings. In this process a feedstock powder is heated by a hot gas jet to a full- or semi-molten state and simultaneously accelerated to supersonic velocity toward a substrate. Upon impact on the substrate surface, thin and circular splats are formed and rapidly solidify. By accumulating these splats on the substrate, the coating can be fabricated. While HVOF spray process can deposit dense WC-Co coatings, the mechanical properties of HVOF sprayed coatings are still inferior to sintered bulk materials. For example, while fracture toughness of as-sprayed WC-Co coatings are about 3~5 $MPam^{1/2}$, the bulk material has 15~30 $MPam^{1/2}$ depending on the volume fraction of binder phase and the carbide size[1-6]. HVOF-sprayed WC-Co coating suffers from high temperature leading to decarburization and dissolution of carbides during deposition and resulting in formation of brittle W_2C and Co-W-C (η) phase. These brittle phases are responsible for their inferior mechanical properties[3,7,8].

In recent years, cold spray method attracts wide interest. In this process, coatings of ductile materials can be produced without significant heating of the sprayed powders. Unmelted particles are accelerated to velocities in the order of 600~1000m/s and bond with a substrate after impact by kinetic energy. Kim et.al.[9,10] fabricated WC-12Co by this method and obtained extremely hard coatings (1926~2127Hv). Although the overall performances of these WC-Co coatings such as toughness and wear resistance are still unknown, it clearly indicates the capability to fabricate the coatings without detrimental phases such as W_2C and η phase, and the possibility to develop new coatings which consist of the similar microstructure and mechanical properties of sintered bulk materials.

Recently Kawakita et al.[11] have developed the improved HVOF process named Warm Spray and succeeded to fabricate dense, thick and clean titanium coatings in the atmosphere. The schematic of the Warm Spray process is shown in Fig.1, which has a mixing chamber between a combustion chamber and a powder feed port in order to control the combustion gas temperature with nitrogen gas. With this new process, the in-flight particle temperature can be successfully controlled and the coatings deposited under the temperature lower than the melting point of the powder are possible. Compared to the cold spray process, in which the available highest gas temperature is approximately 1173K[12] due to the direct gas heating, the warm sprayed process provides a capability to control the particle temperature in the range of 1000~2000K by adjusting the amount of fuel, oxygen and nitrogen gas induced into the combustion chamber. This is the most advantageous feature of the Warm Spray deposition because it provides a possibility to deposit a various materials with their suitable process temperatures. Thus, WC-Co can be sprayed under lower temperature than the eutectic of WC and Co system. This means the powder keeps solid state during spraying. Based on numerical simulations of solid particle impacts[12-14], better bonding between particles can be expected with higher particle temperature. Therefore,

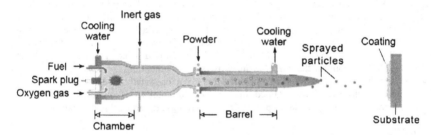

Figure 1. Schematic of Warm Spray (Improved HVOF) deposition.

Table I. Spray conditions for WC-Co coatings by Warm Spray deposition.

Sample ID	A	B	C	D	E	F
Barrel length (mm)	101.6	101.6	101.6	203.2	203.2	203.2
Spray distance (mm)	150	200	200	150	200	200
N$_2$ flow rate (l/min)	500	500	750	750	500	750
Fuel (L/min)	0.32					
Oxygen (L/min)	713					

higher temperature capability of Warm Spray technique is obvious advantage over Cold Spray process. In the present work, Warm Spray deposition was applied to fabricate *degradation free* WC-12Co coatings. The microstructure and the mechanical properties were investigated for samples under various spray conditions.

EXPERIMENT

Commercial WC-12wt.%Co powders were sprayed on carbon steel (JIS: SS400) substrates by Warm Spray equipment. The barrel length, spray distance, and nitrogen flow rate were varied in order to study the effects of these process parameters onto a coating microstructure and mechanical properties such as hardness, Young's modulus, and fracture toughness. The details of the spray conditions were summarized in Table I.

The crystal structure and phases of the feedstock powder and the coatings were analyzed by an X-ray diffractometer. Cross-sectional microstructure was examined by Scanning Electron Microscope (SEM).

Vickers microhardness tests were carried out with a load of 3N and dwell time of 15s on polished cross-section. Fracture toughness, K_{IC}, was evaluated by indentation method with a load of 98N and dwell time of 15s under an assumption that cracks generated from indentation were Palmqvist type. K_{IC} can be given by,

$$K_{IC} = 0.0193 \left(H_V D \right) \left(E / H_V \right)^{2/5} \left(c \right)^{-1/2} \tag{1}$$

where Hv is the Vickers hardness, E is the Young's modulus, D is the half-diagonal of the indentation and c is the indentation crack length[2]. The Young's moduli of coatings were evaluated from the gradient of a load-displacement plot at the onset of unloading by the indentation technique[15] using dynamic ultra microhardness tester (DUH-201, Shimadzu Corporation). At least five indentations were made for each sample. In each case the indentation conditions were identical with a maximum load of 120mN and a loading rate of 4.4mN/sec.

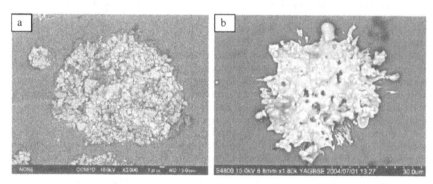

Figure 2. BSE images of splats deposited by (a) Warm Spray deposition and (b) conventional HVOF.

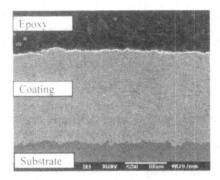

Figure 3. WC-Co coating deposited by warm spray under condition D.

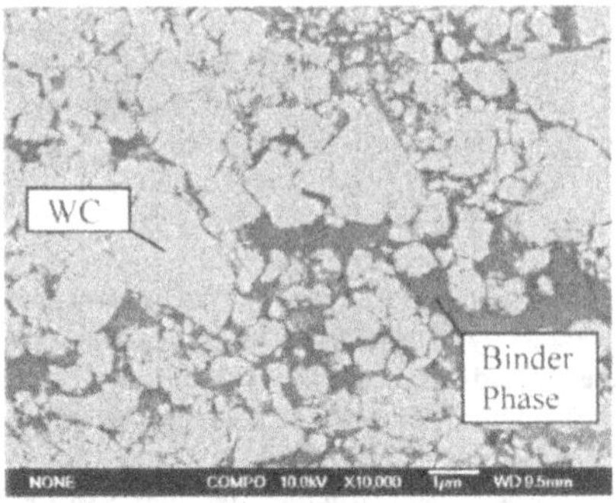

Figure 4. Higher magnification image of the coating microstructure (condition F).

RESULTS AND DISCUSSION

Splat Microstructure

A backscattered electron (BSE) image of a WC-Co splat deposited by the Warm Spray process is shown in Fig. 2a. A splat image of conventional HVOF spraying is also shown for a comparison[1] (Fig. 2b). While the HVOF splat indicates a formation of liquid phase during spraying process, the splat formed by the Warm Spray shows blocky microstructure of carbides and no sign of melting implying that the particle temperature kept lower than the eutectic of WC and Co system and thus the solid particle impacted on the substrate surface. The bonding mechanism by solid particle impact is still unclear. For a metal case, one of the proposed bonding mechanisms is the occurrence of the adiabatic shear instability at interface due to high speed impact and high pressure. For WC-Co case, plastic deformation of Co binder at impact may be the key but the details remains to be studied in future.

Coating Microstructure

Figure 3 shows the cross sectional image of the fabricated coating under condition D. By applying the Warm Spray process, a coating with a thickness of 150 ~ 200μm was successfully fabricated. The coating microstructure was dense and contained substantially higher amount of WC compared with conventional HVOF sprayed coatings probably due to the reduction of the detrimental reactions such as decarburization and dissolution of carbides during process (Fig. 4). The angular morphology of carbides can be also recognized.

Among the spray condition A~F, the condition B and C could not obtain WC-Co coatings successfully. For condition B, the coating could be deposited on the substrate but it delaminated during machining. In the case of condition C, the coating was delaminated during the cooling period after spraying. Apparently, samples of condition B and C had weak bonding strength with the substrates. Except for B and C, the coatings could be deposited. Comparison of condition B, C, E, F suggests that the barrel length which mainly affects particle flying velocity has large effects and longer barrel, that is, higher particle velocity is preferable to fabricate coatings by Warm Spray deposition. For Cold Spray deposition, the concept of the critical velocity over which particle begins to attach on the substrate surface has been proposed by experiments and simulations[12-14,16,17]. Although more detail experiments are necessary, the present experiment ensures the existence of the critical conditions to cause successful bonding of a WC-Co particle and a substrate.

XRD analysis

Figure 5 shows examples of the XRD patterns of the powder and the coating. Both the original powder and the coating show that they are only consisted of WC and Co. Although a broadening of Co peaks in the coating can be recognized at around 43°. Thus, it can infer that the

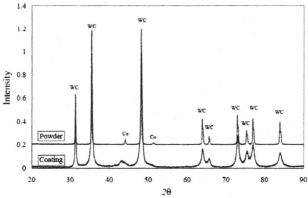

Figure 5. A Comparison of the XRD patterns of the original feed stock powder and the coating.

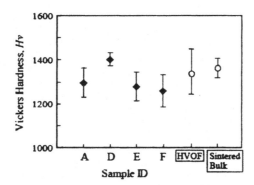

Figure 6 Variations of Vickers hardness for spray conditions and the comparison with the sintered bulk WC-12Co and HVOF sprayed coating results[3].

deposited coatings are essentially degradation free coatings without detrimental phases such as W_2C and η phase and it has quite similar microstructure to the sintered bulk WC-Co at least in terms of chemical compositions. The reason of the broadening of Co peaks is unclear. A possible explanation might be a significant amount of work hardening on the Co binder generated during the particle impact.

Mechanical Properties

Figure 6 shows a plot of hardness values of the coatings. The hardness values of the sintered bulk and the conventional HVOF sprayed WC-12Co coatings are also shown for a comparison[3,18]. The hardness of all samples were in the range of 1200~1400Hv. Except for the

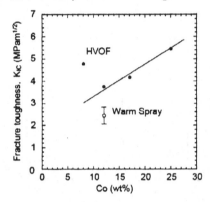

Figure 7. A comparison of the fracture toughness of the coatings deposited by HVOF and Warm Spray.

Figure 8. Defects in the coating

sample D, the hardness of Warm Sprayed coatings is slightly lower than those of the HVOF and bulk material, but it can be said that the difference is small and almost comparable.

However, the fracture toughness of Warm Sprayed coatings was found to be lower than the conventional HVOF data obtained in the previous work as plotted in Fig.7[3]. While the toughness of Warm Sprayed coatings are in the range of 2~3 MPam$^{1/2}$, it is ~4 MPam$^{1/2}$ for the HVOF coatings. Toughness of the sintered bulk of WC-12Co is 12~18 MPam$^{1/2}$ and there exists significant difference. As observed in Fig. 8, there are considerable amounts of pores in the coating and also some gaps between carbides. These defects can be the reason why the fabricated coatings have lower toughness than expected. The current new process is based on an impact between a solid particle and a solid substrate. The further investigations and optimizations are necessary to develop fully dense degradation-free cermet coatings by this new Warm Spray deposition.

CONCLUSIONS

WC-12wt%Co coatings could be successfully deposited by newly developed Warm Spray (improved HVOF) process which can control the temperature of sprayed particles and keep them solid state without melting. The obtained coatings consist of only WC and Co, and essentially degradation-free microstructure could be retained in the spray process.

The mechanical properties of the coatings were evaluated by measuring Vickers hardness, Young's modulus, and fracture toughness. Those values are compared with the data of HVOF sprayed coatings and sintered bulk material. The hardness was almost comparable but fracture toughness was lower than other two processes. Although further investigation of microstructure and optimization of spray conditions are required in future, the present study clearly demonstrated the potential of Warm Spray process.

REFERENCES

[1]M. Watanabe, A. Owada, S. Kuroda, and Y. Gotoh, "Effect of WC size on interface fracture toughness of WC-Co HVOF sprayed coatings," *Surf. Coat. Technol.*, **201** (3-4), 619-627 (2006).

[2]M. M. Lima, C. Godoy, J. C. Avelar-Batista, and P. J. Modenesi, "Toughness evaluation of HVOFWC-Co coatings using non-linear regression analysis," *Mater. Sci. Eng. A-Struct. Mater. Prop. Microstruct. Process.*, **357** (1-2), 337-345 (2003).

[3]P. Chivavibul, M. Watanabe, S. Kuroda, and K. Shinoda, "Effects of carbide size and Co content on the microstructure and mechanical properties of HVOF-sprayed WC-Co coatings," *to be submitted,* (2007).

[4]V. D. Krstic and M. Komac, "Toughening in Wc-Co Composites," *Philos. Mag. A-Phys. Condens. Matter Struct. Defect Mech. Prop.*, **51** (2), 191-203 (1985).

[5]M. Nakamura and J. Gurland, "Fracture-Toughness of Wc-Co 2-Phase Alloys - Preliminary Model," *Metallurgical Transactions a-Physical Metallurgy and Materials Science,* **11** (1), 141-146 (1980).

[6]J. L. Chermant and F. Osterstock, "Fracture Toughness and Fracture of Wc-Co Composites," *J. Mater. Sci.,* **11** (10), 1939-1951 (1976).

[7]J. M. Guilemany, J. M. de Paco, J. Nutting, and J. R. Miguel, "Characterization of the W2C phase formed during the high velocity oxygen fuel spraying of a WC+12 pct Co powder," *Metall. Mater. Trans. A-Phys. Metall. Mater. Sci.,* **30** (8), 1913-1921 (1999).

[8]C. Verdon, A. Karimi, and J. L. Martin, "A study of high velocity oxy-fuel thermally sprayed tungsten carbide based coatings. Part 1: Microstructures," *Mater. Sci. Eng. A-Struct. Mater. Prop. Microstruct. Process.*, **246** (1-2), 11-24 (1998).

[9]H. J. Kim, C. H. Lee, and S. Y. Hwang, "Superhard nano WC-12%Co coating by cold spray deposition," *Mater. Sci. Eng. A-Struct. Mater. Prop. Microstruct. Process.*, **391** (1-2), 243-248 (2005).

[10]H. J. Kim, C. H. Lee, and S. Y. Hwang, "Fabrication of WC-Co coatings by cold spray deposition," *Surf. Coat. Technol.*, **191** (2-3), 335-340 (2005).

[11]J. Kawakita, S. Kuroda, T. Fukushima, H. Katanoda, K. Matsuo, and H. Fukanuma, "Dense titanium coatings by modified HVOF spraying," *Surf. Coat. Technol.*, **201** (3-4), 1250-1255 (2006).

[12]T. Schmidt, F. Gärtner, H. Assadi, and H. Kreye, "Development of a generalized parameter window for cold spray deposition," *Acta Mater.*, **54** (3), 729-742 (2006).

[13]H. Assadi, F. Gärtner, T. Stoltenhoff, and H. Kreye, "Bonding mechanism in cold gas spraying," *Acta Mater.*, **51** (15), 4379-4394 (2003).

[14]K. Yokoyama, M. Watanabe, S. Kuroda, Y. Gotoh, T. Schmidt, and F. Gärtner, "Simulation of solid particle impact behavior for spray processes," *Mater. Trans.*, **47** (7), 1697-1702 (2006).

[15]J. A. Thompson and T. W. Clyne, "The effect of heat treatment on the stiffness of zirconia top coats in plasma-sprayed TBCs," *Acta Mater.*, **49** (9), 1565-1575 (2001).

[16]C. J. Li, W. Y. Li, and H. L. Liao, "Examination of the critical velocity for deposition of particles in cold spraying," *J. Therm. Spray Technol.*, **15** (2), 212-222 (2006).

[17]T. Marrocco, D. G. McCartney, P. H. Shipway, and A. J. Sturgeon, "Production of titanium deposits by cold-gas dynamic spray: Numerical modeling and experimental characterization," *J. Therm. Spray Technol.*, **15** (2), 263-272 (2006).

[18]G. S. Upadhyaya, *Cemented Tungsten Carbides. Production, Properties, and Testing.* (Noyes, 1998).

FRACTURE MECHANICS ANALYSIS OF COATINGS UNDER CONTACT LOAD

Yumei Bao, Guozhong Chai, Weina Hao

The MOE Key Laboratory of Mechanical Manufacture and Automation
Zhejiang University of Technology
Hangzhou 310032, P.R.China

ABSTRACT
With better temperature, friction, and wear resistant properties than their substrate counter parts, coatings are increasingly used in a large number of industries that include the aerospace, automobile, computer, machining, and precision manufacturing, either to extend the fatigue life of mechanical components in contact or to provide low friction coefficients and wear factors. Because of the difference of the elastic and thermal properties between the coating and substrate, the mechanical performance of coated materials is different from those isotropic homogeneous materials under loadings. Even under the same loadings, the stresses produced in coated materials may be larger than in isotropic homogeneous ones, so it is easier for the cracks to initiate and grow, and eventually the failure takes place. Especially in the weak interfacial of the coating and substrate, the cracks initiate and grow easily due to the large stresses, resulting in debonding and rupture. The coating/ substrate system with predetermined cracks exerted by a cylindrical indenter is studied with ANSYS program, the stress intensity factors of the interfacial cracks and the surface cracks are analyzed under different ratio of Young's modulus of the coating/ substrate, crack length, and crack location. The research is helpful for mechanical analysis of coating/ substrate system and design of coating material.

INTRODUCTION
Coatings are increasingly used in a large number of industries such as the aerospace, automobile, computer, machining, and precision manufacturing, either to extend the fatigue life of mechanical components in contact or to provide low friction coefficients and wear factors. The typical application is in tribological systems to modify surface properties of engineering materials and components. The performance of the tribological system can be improved greatly and the life of the tribological parts can be extended due to the coatings. This is because surface properties, such as surface hardness, wear resistance, degradation resistance, and corrosion resistance of tribological components, can be considerably improved by effectively using coatings, the typical examples are hard coatings of TiN and TiCN used extensively in tool industries to increase service time of cutting tools, dies, and punches and WC and diamond like carbon used in bearings, gears, etc. to minimize wear. While soft coatings of MoS_2 and graphite, on the other hand, are used as solid lubricants [1-4].

There are many examples of applications of coatings where a mechanical contact of the coated body with a counterpart is formed and the mechanical interaction is unavoidable. Response of coated components to mechanical loads depends on not only the properties of the coating and the substrate, but also the interactions between them [5]. Coating failure such as cracking and debonding may occur due to overstressing. The cracks propagate, or new cracks form faster than homogeneous materials under loading [6-7].

Therefore the bond strength of the coating/substrate is one of the important properties of the coating system, but because of the difference of the elastic and thermal properties between the coating and substrate, the mechanical performance of coated materials is different from those isotropic homogeneous materials under loadings, and it is very difficult to obtain precise analytical solution on the contact problem of the coating/substrate system. In order to obtain a more detailed knowledge of the contact behavior of the coated surfaces, numerical modeling techniques are often employed.

In this paper, the coating/ substrate system with predetermined cracks exerted by a cylindrical indenter is analyzed with ANSYS program, the stress intensity factors of the interfacial cracks and the surface cracks are analyzed under different ratio of Young's modulus of the coating/ substrate, crack length, and crack location.

FEM MODEL
The surface and interfacial crack is shown in Fig.1, exerted by a cylindrical indenter.

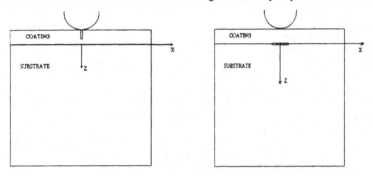

Fig.1 Surface and interfacial crack under cylindrical indenter

The coating/ substrate composite is considered to be perfectly combined therefore they can be regarded as a whole. Assume the width of the substrate 200mm, the height 200mm – the thickness of the coating (t) 1mm. The radius of the cylindrical indenter is 10mm. The Young's modulus and Poisson's ratio of the substrate are E_s=210GPa and μ_s =0.3. The half crack length is denoted by c.

Numerical modeling is done by the ANSYS finite element program. The crack is simulated by two superposed lines, the tip is divided into 16 six-noded isoparametric trilateral elements, and eight-noded isoparametric quadrilateral elements are used for the solution model domain (see Fig. 2). The coordinates are located at the interface of the coating and the substrate, with the origin located in the center of the contacted zone, as shown in Fig.1.The mesh is refined in the contact zone to ensure the precision in view of the large stress gradient. The coating/ substrate is divided into 10 equal parts when $Z \leq t$ and 40 equal parts when $|X| \leq 10t$ to get denser mesh in the contact zone. The total numbers of the elements are 31433.

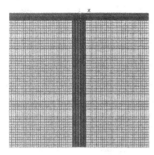

Mesh of the crack tip

Fig.2 FEM mesh

In the following analysis, the elastic modulus ratio between the coating and the substrate Ec/Es is 1,5 and 10 respectively. The dimensionless factor $\overline{K}_I$ is defined as $\overline{K}_I = K_I / K^*$, where $K^* = P\sqrt{\pi \cdot /t}$ (in which P is the pressure exerted).

RESULTS AND DISCUSSIONS
Under cylindrical indenter, the variation of the stress intensity factor of the surface and interfacial crack with the crack length and its location is studied.

Surface Crack
Fig.3 shows that the value of $\overline{K}_I$ decreases with the crack length, when the surface crack is located in the center of the contact area. It is clear that the larger E_c / E_s ratio results in the larger $\overline{K}_I$.

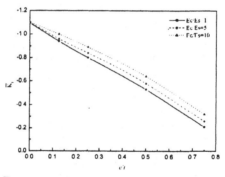

Fig.3 $\overline{K}_I$ of the surface crack tip varying with the crack length

The value of $\overline{K}_I$ when the crack length is 0 is computed by the following empirical equation:

$$\overline{K} = \lim_{a \to 0} \frac{1.12\sigma_x \sqrt{\pi a}}{P\sqrt{\pi a}/t} = 1.12\sigma_x t/P$$

where σ_x is the stress at the origin of coordinates.

Fig. 4 shows how $\overline{K}_I$ of the surface crack varied with the location of the crack, under different E_c/E_s ratio and crack length, when the surface crack is moving away from the center of the contact area.

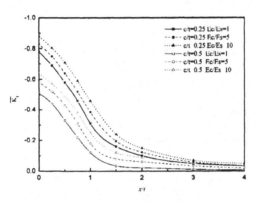

Fig. 4 $\overline{K}_I$ of the surface crack tip varying with the crack location
(x is the distance between the crack and the center)

It is clear that the $\overline{K}_I$ value decreases with the x, and when the distance between the crack and the symmetrical axis is $4t$, $\overline{K}_I$ approaches to 0. The smaller ratio of E_c/E_s results in smaller magnitude of $\overline{K}_I$. While the longer the crack, the smaller the absolute $\overline{K}_I$.

Interfacial Crack
Currently, the interfacial crack stress intensity factor can not be obtained directly from any finite element software, for crack field is made up of different materials. In what follows, the stress field of crack tip and the related displacement near crack tip of corresponding point are computed with ANSYS, and then the stress intensity factor is calculated with a series of equations (proposed by Muskhelishviliv [8]). Results of the bimaterial plate with a central interfacial crack are presented to confirm the accuracy of the above method, considering different ratio of Young's Modulus of coating and substrate [9]_ see Fig. 5 and Tab.1 _.

Table 1 $\overline{K}_I$ * of bimaterial rectangular plate with a central interfacial crack

E_1/E_2	Numerical solution obtained	Results of literature [9]
1	1.000	1.000
3	0.994	0.988
5	0.971	0.968
10	0.955	0.953
100	0.954	0.952

* : $\overline{K}_I = K_I / K^*$ _ $K^* = P\sqrt{\pi a}$

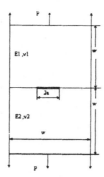

Fig 5 bimaterial plate with a central interfacial crack

Fig 6 shows how the dimensionless $\overline{K}_I$ varies with the crack length under different E_c / E_s, when the crack is locating at the center of the contact area. When the crack length c=0, $\overline{K}_I = \sigma_{.t} / P$. Considering the symmetry of the load and geometry, $\overline{K}_I$ of the left tip and the right tip of the interfacial crack are the same. It is clear that $\overline{K}_I$ decreases with the crack length, and the effect of E_c / E_s on $\overline{K}_I$ is smaller and smaller when the crack is longer and longer. The smaller E_c / E_s, the larger magnitude of $\overline{K}_I$.

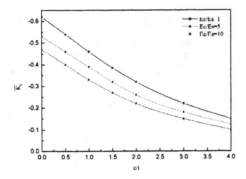

Fig. 6 $\overline{K}_I$ of the interfacial crack tip varying with the crack length

Fig 7 and 8 shows how the dimensionless $\overline{K}_I$ of the left and right tip of the interfacial crack varies with the crack length under different E_c / E_s, when the crack is moving away from the center of the contact area. It is clear that the variation of the left and right tip of the crack is the same on the whole. $\overline{K}_I$ of the left and right tip decreases when the crack is moving away from the center, but the decreased value of the right tip is larger than that of the left one. And the absolute $\overline{K}_I$ is smaller if the crack is longer and the more obvious the decreasing trend is shown. In addition the smaller E_c / E_s, the larger magnitude of $\overline{K}_I$ is.

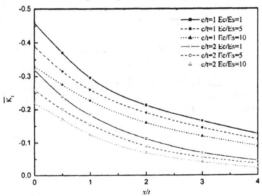

Fig. 7 $\overline{K}_I$ of the left tip of the interfacial crack varying with the crack location

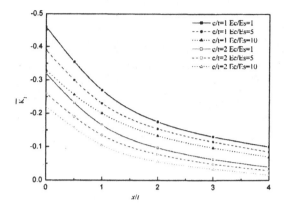

Fig. 8 $\overline{K}_I$ of the right tip of the interfacial crack varying with the crack location

CONCLUSIONS

As for either the surface crack or the interfacial crack, the ratio of Young's modulus between coating and substrate (E_c / E_s) has obvious influence on the stress intensity factor $\overline{K}_I$. The larger ratio of Young's modulus between coating and substrate (E_c / E_s) results in smaller magnitude of $\overline{K}_I$ for surface crack and larger magnitude of $\overline{K}_I$ for interfacial crack. While the longer crack makes for the smaller magnitude of $\overline{K}_I$. And the absolute $\overline{K}_I$ decreases continuously when the crack is moving away from the center of the contact area. As for the interfacial crack, $\overline{K}_I$ of the left tip is larger than that of the right one.

The proposed efforts can provide foundations for the design and preparation of coated materials, especially for the research on the crack growth and even failure in the coating system under mechanical contact and other loads.

ACKNOWLEDGEMENTS

The presented research work was financially supported by National Natural Science Foundation of China (50375145) and Zhejiang Natural Science Foundation No. Y104279 and Y604051

Contact E-mail address: baoym@zjut.edu.cn (Yumei Bao)

REFERENCES

[1] Gian Mario Bragallini, Maria Pia Cavatorta, Philippe Sainsot. Coated contacts: a strain approach. Tribology International. 2003,36: 935–941

[2] K. Kaddouri, M.belhouari, et al. Finite Element Analysis of Crack Perpendicular to Bi-material Interfacial: Case of Couple Ceramic-metal. Computational Materials Science 2006, 35:53~60

[3] A.A.Zalounima, J.H.Andreasen. A Crack Coated Solid Subjected to Contact Loading. Wear 2004,257,671~686

[4] K.Aslantas, S.Tasgetiren. Debonding between Coating and Substrate due to Rolling-sliding Contact. Materials and Designs $-$ 2002,23,571-57

[5] N. SchwarzerU. Coating design due to analytical modelling of mechanical contact problems on multilayer systems. Surface and Coatings Technology, 2000, 133-134: 397-402

[6] K. Aslantas, S. Tasgetiren. Debonding between coating and substrate due to rolling–sliding contact. Materials and Design, 2002, 23: 571–576

[7] Leon L. Shaw1, Brent Barber, et al. Measurements of the interfacial fracture energy of thermal barrier coatings. Scripta Materialia, 1998, 39(10): 1427-1434

[8] Muskhclishvili N I. Some basic problems of the mathematical theory of elasticity. 4th ed., Noordhoff Groningen,1972

[9] Yuqiu Long. Introduction to advanced finite element. 1st ed., Beijing $-$ Tshinghua University Press, 1992

Coatings for
Space Applications

HEAT TREATMENT OF PLASMA-SPRAYED ALUMINA: EVOLUTION OF MICROSTRUCTURE AND OPTICAL PROPERTIES

Keith S. Caruso[*a], David G. Drewry[a], Don E. King[a], Justin S. Jones[b]

[a]The Johns Hopkins University Applied Physics Laboratory, 11100 Johns Hopkins Rd, Laurel, MD USA 20723-6099
[b]The Johns Hopkins University Whiting School of Engineering, 3400 North Charles Street, Baltimore, MD 21218-2681

ABSTRACT

Research conducted by The Johns Hopkins University related to space exploration has involved design of optical ceramic coatings for thermal management in near-solar environments. One such coating being investigated is alumina (Al_2O_3), plasma sprayed onto a carbon-carbon (C-C) conical heat shield as a concept envisioned for NASA's Solar Probe mission. A thin Al_2O_3 coating applied to C-C reflects the majority of the visible and near infrared (NIR) solar irradiance, while both materials re-emit absorbed infrared energy. Optical test data for Al_2O_3 coated C-C coupons show that the solar-absorptance-to-IR emittance ratio (α_S/ϵ_{IR}) is less than 0.6. Compared to an uncoated carbon-carbon heat shield, the coated version reduces the predicted heat shield temperature by over 250K at a distance of 4 solar radii ($4R_s$) from the center of the sun. The feasibility of the Solar Probe mission is dependent on this reduction in temperature resulting from control of optical properties.

Plasma-sprayed Al_2O_3 deposits (splats) consist of several metastable crystallographic modifications (phases), which transform to the stable α-phase during heat treatment at elevated temperatures. Heat treatments of Al_2O_3 coatings at critical phase transformation temperatures were performed to study microstructural and morphological changes. During heat treatment of the plasma-sprayed Al_2O_3 coatings, there were changes in porosity, density, crystal structure, and surface texture, as well as significant grain growth at temperatures above 1180°C. These phenomena have a significant effect on optical properties of the ceramic coatings, as shown by spectroscopic measurements after each heat treatment. Specifically, there were significant increases in UV, visual, and NIR reflectivity and changes in thermal band absorption due to the heat treatment schedules applied. Such changes in optical properties could have significant impact on the Solar Probe heat shield equilibrium temperature during solar approach. Along a mission trajectory, the predicted temperature of an Al_2O_3 coated C-C heat shield undergoing in-situ phase transformations and crystallization was compared to that of a heat shield stabilized to α-phase Al_2O_3 prior to launch.

INTRODUCTION

Solar Probe will be an historic mission, flying into one of the last unexplored regions of the solar system, the Sun's corona, for the first time. Approaching as close as 3 solar radii (R_S) above the Sun's surface, Solar Probe will employ a combination of in-situ measurements and imaging to achieve the mission's primary scientific goal: to understand how the Sun's corona is heated and

177

how the solar wind is accelerated. Solar Probe will revolutionize our knowledge of the physics of the origin and evolution of the solar wind. Moreover, by making the only direct, in-situ measurements of the region where some of the deadliest solar energetic particles are energized, Solar Probe will make unique and fundamental contributions to our ability to characterize and forecast the radiation environment in which future space explorers will work and live[1].

The baseline Solar Probe mission includes two flybys of the Sun, separated by 4.6 years. The timing allows scientific measurements of the solar wind and corona to be made at different phases of the 11-year solar cycle, independent of the launch date. Solar Probe will use a Jupiter gravity assist to achieve a polar orbit about the Sun with a perihelion of $4R_S$, that is, 3 R_S above the Sun's surface as shown in Figure 1. The Solar Probe spacecraft is designed to both support and protect the science payload over the extreme environmental variations that it will experience during the mission. The notional spacecraft design includes a thermal protection system (TPS) in front of the science payload. The TPS consists of three parts: the conical primary shield, whose purpose is to limit the temperature and reject most of the solar heat to space; the secondary shield, which reduces heat transfer to the instrument bus to a reasonable level; and the struts, which support the TPS shields at a safe distance from the spacecraft bus (Figure 1).

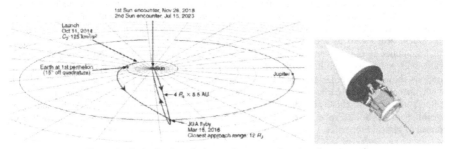

Figure 1. Solar Probe mission trajectory and notional spacecraft

The primary shield provides the first step in solar protection by reflecting and emitting the majority of the incident flux away from the observatory (Figure 2). It reduces the solar heat by 75% by minimizing its view factor to the Sun and maximizing the view to space. A 40% reduction of the remaining input is achieved through use of a "white" ceramic coating with high solar reflectivity (ρ_S) and high infrared emissivity (ε_{IR}) over the carbon-carbon conical structure. The baseline ceramic coating to reduce the heat shield temperature via control of optical properties is alumina (Al_2O_3). The integrity of the coating and its optical properties at high temperatures are of utmost importance for mission success. Therefore, this study is focused on Al_2O_3 coatings, air plasma-sprayed onto carbon-carbon substrates, and their properties during heat treatment to mission-like temperatures.

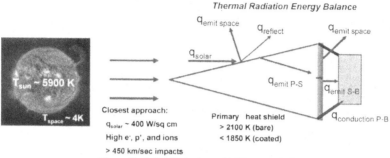

Figure 2. Solar Probe heat shield energy balance

   The incorporation of optical coatings onto the primary shield provides a passive thermal management approach designed to reduce the primary shield equilibrium temperature. The substrate material, C-C, inherently absorbs visible energy because of its black color, but is also a strong emitter of infrared energy. As shown in Figure 3, the peak levels of solar irradiance are located in the visible region, but the peak energy radiated by a blackbody is in the infrared. Blackbody emission shifts toward lower wavelengths as the temperature rises (ie, as Solar Probe approaches the Sun). Ideally, a coating would reflect a majority of energy in the visible wavelengths and then either emit or be transparent in the IR region so that the IR emittance of the C-C can be used. The concept of an ideal coating on the C-C substrate is depicted in Figure 4. Ultimately, the heat shield temperature is determined by the ratio of solar absorptance (α_s) to infrared emittance (ε_{IR}) as shown in equation 1 below. Solar absorptance is the weighted average of absorptivity (one minus the reflectivity) with wavelength over the solar spectrum (equation 2). Infrared emittance is the weighted average of emissivity with wavelength over the blackbody energy curve (equation 3).

$$Temperature_{HeatShield} = Temperature_{Blackbody}\left(\frac{\alpha_s}{\varepsilon_{IR}}\right)^{0.25} \qquad \text{(eq. 1)}$$

$$\alpha_s = 1 - \rho_s = \frac{\left(\sum_i^n \rho(\lambda_i)E(\lambda_i)\Delta\lambda_i\right)}{\left(\sum_i^n E(\lambda_i)\Delta\lambda_i\right)} \qquad \text{(eq. 2)}$$

$$\varepsilon_{IR} = 1 - \rho_{IR} = \frac{\left(\sum_i^n \rho(\lambda_i)E(\lambda_i)\Delta\lambda_i\right)}{\left(\sum_i^n E(\lambda_i)\Delta\lambda_i\right)} \qquad \text{(eq. 3)}$$

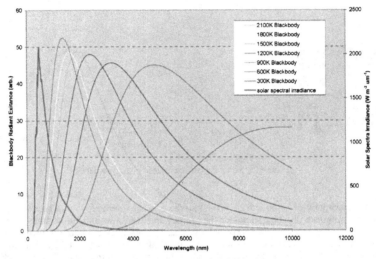

Figure 3. Solar spectrum and Planck blackbody curve

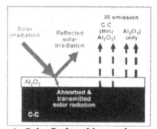

Figure 4. Solar Probe white coating concept

Literature surveys and test data revealed the mission-enabling potential of white ceramic coatings for thermal control of the Solar Probe primary shield near the Sun[2]. To characterize these coatings, optical data (as a function of temperature and wavelength) were collected over the anticipated exo-atmospheric solar spectrum by using both discrete wavelength lasers and a continuous wavelength spectrophotometer. The discrete lasers were selected to replicate significant portions of the solar spectrum and to capture data in the near- to mid-IR band. These lasers were used to measure test coupons both at room temperature and at elevated temperatures. The spectrophotometer (which operated only at room temperature) enabled validation of the discrete laser results and characterization of the materials over the full range of wavelengths from the UV through the IR. Continuous data for Al_2O_3 and Pyrolytic Boron Nitride (PBN) coatings on C-C taken with the spectrophotometer are shown in Figure 5, along with the solar spectrum and the 2100K Planck blackbody curve. Both "white" coating systems strongly reflect solar energy, then begin to increasingly absorb (and emit) at higher wavelengths in the infrared. This trend results in a α_S/ϵ_{IR} ratio less than unity ($\alpha_S/\epsilon_{IR} \approx 0.6$ for Al_2O_3 and $\alpha_S/\epsilon_{IR} \approx 0.45$ for PBN), and therefore, reduced heat shield temperatures compared to bare C-C with no ceramic coating ($\alpha_S/\epsilon_{IR} \approx 1$). Test data were incorporated into a spectral-based radiant thermal model to accurately predict the spacecraft's primary shield temperature as a function of distance from the Sun[3].

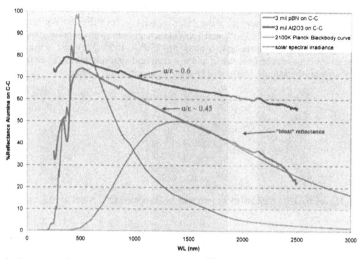

Figure 5. Reflectance of coatings plotted with the solar spectrum and 2100K blackbody curve

Several authors have studied phase changes and microstructural development of plasma-sprayed alumina coatings. Stumpf et al.[4] identified the sequence of phase transitions in alumina by x-ray diffraction; this sequence has since become generally accepted. Petersson et al.[5] describe the plasma spraying process, why alumina precipitates from the melt as gamma (γ) phase material, and how it transforms via delta (δ) phase to the stable alpha (α) phase at temperatures above 1100°C. McPherson[6] studied the splat formation process in plasma spraying of alumina and the formation of mostly γ-phase during deposition. Damani and Wanner[7] studied the microstructural development of plasma-sprayed alumina as a function of annealing temperature. They showed by x-ray diffraction the phase changes from deposited metastable γ-phase to δ-phase then θ-phase then the stable α-phase at 1180°C. Porosimetry measurements showed little change in porosity until the transformation to α-phase material, resulting in an increase of approximately 4% porosity. This study monitors the changes in optical properties (α_S and ϵ_{IR}) of plasma-sprayed alumina coatings on C-C substrates during phase changes and microstructural development due to heat treatments similar to those performed by the above authors. These optical properties of the ceramic coating are of utmost importance to the Solar Probe mission for reasons outlined above.

EXPERIMENTAL

All material samples were made at the Johns Hopkins University Advanced Technology Lab using an air plasma spray system. Alumina coatings approximately 0.010" thick were applied to fully densified C-C panels (Figure 6), then cut into smaller pieces using a diamond saw. Heat treatments were performed in an alumina tube furnace (Figure 7) with inert gas purge to avoid oxidation of the C-C substrate. Treatment temperatures of 900°C, 1050°C, and 1180° were chosen because they are critical phase transition temperatures of plasma-sprayed alumina from the metastable deposited γ-phase material to the stable α-phase, as identified by Weferes and Misra[8]. Heat treatment temperatures of 1380°C and 1550°C were chosen to study grain growth and sintering of α-phase Al_2O_3 after full conversion to α-phase at 1180°C. 1550°C is close to the predicted temperature of the Solar Probe heat shield and to a common sintering temperature for alumina. For each heat treatment, the sample was at steady-state temperature for 2 hours.

Figure 6. JHU plasma-spray gun and Al_2O_3 coated C-C panel used for testing

Figure 7. Alumina tube furnace used to perform heat treatments of Al_2O_3 coatings

Optical properties of the heat treated coatings were measured using UV/Vis/NIR and infrared spectrometers (Figure 8), both with integrating spheres. Reflection measurements were taken with the Al_2O_3 coatings on C-C substrates so the incident light is either reflected or absorbed, as will generally be the case for the Solar Probe spacecraft. Coatings were then removed from the C-C substrates by oxidation in an air furnace at 800°C for 4 hours, then transmission measurements were taken through the heat treated, free standing coatings. The ultimate properties of interest, α_S and ε_{IR}, were calculated from the continuous spectral data as described in ASTM Standard E903-82[9], using equations 2 and 3 above.

High resolution x-ray diffraction scans of as-sprayed and heat treated coatings were taken to identify phase transitions and correlate these transitions to published literature. SEM images of coating top and bottom surfaces were collected at 2,000x and 15,000x magnification to study microstructural development and grain growth during phase transitions to 1180°C and during further annealing of alpha phase material at 1380°C and 1550°C. Phase transitions and microstructural changes were correlated to changes in reflectivity and transmissivity.

Figure 8. UV/Vis/NIR and infrared spectrometers used to collect reflectance and transmittance data

RESULTS

Phase transitions were identified by taking x-ray diffraction scans of Al_2O_3 coated C-C substrates and comparing those scans to similar results published by other authors. The deposited coating before heat treatment was shown to be primarily γ-phase material. Alpha Al_2O_3 is the only phase stable at all temperatures. However, The metastable γ-phase is produced because γ-Al_2O_3 is more easily nucleated from the melt due to lower interfacial energy between its crystal and the melt. If the rate of cooling is sufficiently rapid, the metastable γ-phase material is retained at ambient temperature. The deposited mixture of γ-Al_2O_3 with cubic structure and α-Al_2O_3 with trigonal structure then transitions to all stable α-phase material during heat treatments at 900°C and above as shown in Figure 9. The coating material transitions from γ- Al_2O_3 to α-Al_2O_3 through the δ- and θ-phases. X-ray patterns showed that after heat treatment at 1180°C for 2 hours, the coating was almost entirely α-Al_2O_3. Above 1180°C the material is stable, but microstructural changes such as terracing and grain growth continue to occur. Figure 10 shows more detail of the x-ray diffraction peaks corresponding to the phase changes. The as-sprayed material has two large γ-phase peaks (at $2\theta = 45°$ and 67°) as well as smaller γ- and α-phase peaks at many locations. The typical ratio of γ- to α-phase in air plasma sprayed Al_2O_3 is 70/30. The scan after 2 hour heat treatment at 900°C shows the beginning of phase transition as the γ-phase peaks begin to split. Phase transitioning is even more apparent after 1050°C annealing, when many δ-phase peaks appear. The scan after 2 hour 1180°C heat treatment shows almost pure α-phase material.

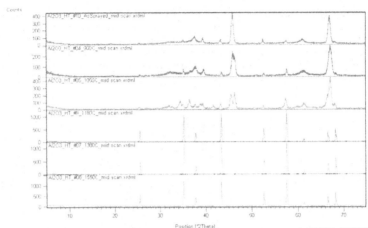

Figure 9. XRD scans of alumina coatings. From top to bottom: as-sprayed, 900°C, 1050°C, 1180°C, 1380°C, and 1550°C heat treatments.

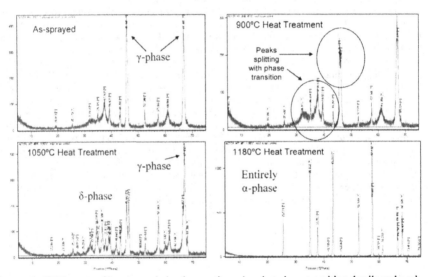

Figure 10. XRD scans of heat treated alumina coatings showing phase transition details and peak locations

SEM images of heat treated coatings were collected to study the changes in microstructure that may affect optical properties of the coating (ie, reflectivity). There are many physical changes that commonly occur during phase transitions, annealing, and sintering that can influence how light interacts with the Al_2O_3 coating material. These changes include crystal structure, porosity (open and closed), density (volume shrinkage), surface roughness, morphological restructuring, and grain growth. SEM images of heat treated coatings provided evidence of several significant morphological changes, confirmed by XRD and optical measurements. Figure 11 shows the splat structure of as-sprayed Al_2O_3 at a magnification of 2,000x, and changes to the structure that occur with each heat treatment. There is little evidence of change at 900°C and 1050°C, but at 1180°C re-crystallization, surface re-ordering, and terracing are apparent. SEM images at 15.000x (Figure 12) show evidence of grain growth beginning at 1180°C and continuing through annealing steps at 1380°C and 1550°C. Grains grow from generally less than a micron in size at 1180°C to several microns after heat treatment at 1550°C. Some fully terraced surfaces and formation of equilibrium crystal shapes are also evident in the SEM micrograph at 1550°C. After 1550°C heat treatment, the original splat structure, and much of the open porosity is lost. Studies by other authors have shown that closed porosity increases after the transition to alpha phase.

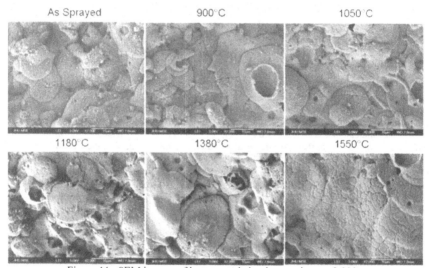

Figure 11. SEM images of heat treated alumina coatings at 2,000x

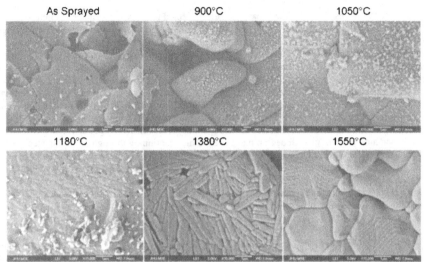

Figure 12. SEM images of heat treated alumina coatings at 15,000x

Phase transitions and microstructural changes were correlated with spectroscopic measurements of the Al_2O_3 coatings. Reflectance from 0.25 to 2.5μm of the as-sprayed and heat treated coatings on C-C are shown in Figure 13 plotted against the solar spectrum, along with the calculated solar absorptance (α_S) of each sample. The peak reflectance, wavelength of peak reflectance, and calculated α_S are listed in Table 1. Peak reflectance increases and shifts towards lower wavelengths during the first three heat treatments at 900°C, 1050°C, and 1180°C. After transition to alpha Al_2O3, reflectivity decreases as grain growth progresses at 1380°C and 1550°C. The same trend is also evident in transmission measurements through the heat treated coatings (Figure 14). Transmittance through the coating decreased due to heat treatments at 900°C, 1050°C, and 1180°C, then increased after further annealing at 1380°C and 1550°C. Increase in transmittance from 1380°C to 1550°C was most significant.

Table 1. Optical property comparison of heat treated coatings

Heat Treatment (°C)	Peak Reflectance (%)	Peak Reflectance Wavelength (nm)	Solar absorptance
As-sprayed	76.1	492	0.29
900	86.5	412	0.21
1050	86.5	412	0.21
1180	91.4	366	0.13
1380	88.5	412	0.16
1550	80.1	336	0.23

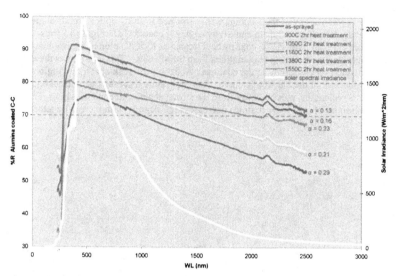

Figure 13. UV/Vis/NIR reflectance of heat treated plasma-sprayed alumina coatings

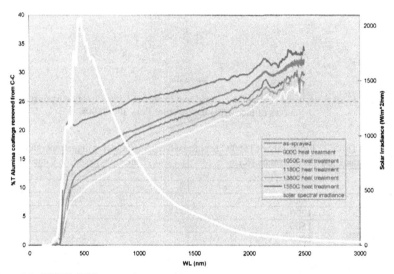

Figure 14. UV/Vis/NIR transmittance of heat treated plasma-sprayed alumina coatings

Infrared spectroscopy of the coatings was performed to study optical trends at higher wavelengths and relate these trends to emissivity of Al_2O_3 coated C-C. Figure 15 shows the reflectance from 2-16μm of an Al_2O_3 coating heat treated at 1550°C, plotted against blackbody energy distributions for increasing temperatures from 300-2100K. This data shows the effect of temperature increase on emissivity of the coatings. At 300K, the peak absorption region of Al_2O_3 is centered over the peak of the blackbody curve. As temperature increases, peak blackbody energy shifts to lower wavelengths, moving away from the strong absorption band of Al_2O_3. This is one reason that emissivity of the Al_2O_3 coatings decreases significantly with increasing temperature (Table 2).

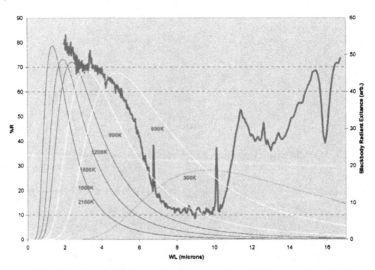

Figure 15. IR reflectance of alumina coating heat treated at 1550°C plotted against blackbody curves of increasing temperature from 300-2100K

Table 2. Emissivity of 1550°C heat treated coating at increasing temperature

Blackbody Temperature (K)	Emissivity of Al_2O_3 Coated C-C
300	0.66
600	0.58
900	0.47
1200	0.40
1500	0.37
1800	0.35
2100	0.33

Figure 16 shows the IR reflectance of all heat treated Al2O3 coatings on C-C versus the 2100K blackbody curve. The reflectance from 2-7μm is lowest for the as-sprayed coating, resulting in the highest IR emissivity (ε_{IR}). Also, coatings annealed at 1380°C and 1550°C lost their strong absorption band from 2.7-3.5μm that is evident in the data for coatings below 1380°C. Increasing heat treatment temperature resulted in higher emissivity as shown in Table 3.

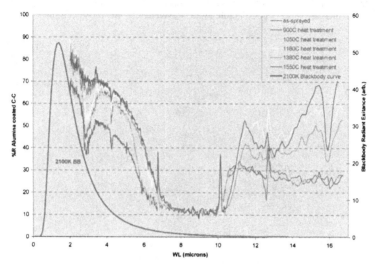

Figure 16. IR reflectance of heat treated alumina coatings plotted against 2100K blackbody

Table 3. Emissivity of heat treated coatings relative to 2100K blackbody

Coating Heat Treatment Temperature (°C)	Emissivity at 2100K
As-sprayed	0.50
900	0.41
1050	0.40
1180	0.35
1380	0.33
1550	0.33

Infrared transmission spectroscopy of the heat treated Al2O3 coatings removed from C-C was performed to study how the absorption characteristics change with microstructure. Figure 17 shows transmittance of each coating material from 2-16μm. The as-sprayed material and coatings annealed to 900°C and 1050°C have similar absorption bands from 2.7-3.5μm and from 5.5-8μm. The Al2O3 coating heat treated at 1180°C is similar to lower temperature treatments below 5.5μm, but above

5.5μm absorption becomes significantly less, most likely due to conversion to alpha phase and the start of grain growth. Absorption in the 1380°C and 1550°C heat treated coatings is significantly less than at lower temperatures as seen in the bands from 2.7-3.5μm and from 5.5-8μm. Transmittance through these coatings heat treated at higher temperatures is increased, which would allow the underlying C-C to more effectively absorb and re-emit infrared energy. Interference effects in the coatings heat treated at 1180°C, 1380°C, and 1550°C are evident at higher wavelengths as Al$_2$O$_3$ grains increase to sizes comparable to the wavelength of the incident light.

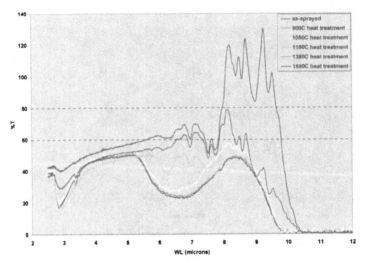

Figure 17. IR transmittance of heat treated alumina coatings

The optical property trends described above significantly impact the Solar Probe mission. The temperature of the heat shield during solar approach is ultimately determined by the ratio of solar absorptance, α_S, to infrared emittance, ε_{IR}, of the Al$_2$O$_3$ coating on C-C substrate system as shown in equation 1. These optical properties are a function of several coating features including crystal structure, porosity, density, surface texture, morphology, and grain size. The spectroscopy data described above show significant changes as physical restructuring of the coatings occurs with heat treatment. In fact, the biggest changes occur when the coating is annealed at 1550°C which is close to the predicted temperature of the heat shield at closest solar approach. Changes in optical properties due to phase change and microstructural evolution, as well as emissivity changes with temperature, must be considered to accurately predict the temperature of the heat shield at any point along the mission trajectory.

Figure 18 shows a comparison of heat shield temperature for C-C with no coating, Al$_2$O$_3$ coating on C-C heat treated to 1550°C to stabilize it prior to the mission, and Al$_2$O$_3$ coating on C-C allowed to heat treat "in-situ" during the mission. A C-C heat shield with no coating has an $\alpha_S/\varepsilon_{IR}$ ratio of approximately 1 along the mission trajectory and will have a temperature similar to a blackbody at any distance from the sun. This is shown by the grey boxes representing temperatures from 300K outside 0.3 Astronomical Units (Au) to 2100K at closest approach (4R$_S$). An Al_2O_3 coated C-C heat shield that is stabilized prior to the mission has a constant α_S, but increases in temperature due to decreasing emissivity as the temperature of the heat shield increases, represented by the blue boxes. An Al_2O_3 coated C-C heat shield that is launched in the as-sprayed (γ-phase) condition has a more interesting trend of temperature increase during the mission, shown by the red boxes. In this scenario, both α_S and ε_{IR} change continuously during solar approach. Solar absorptance decreases dramatically as the coating is heat treated to from 900°C to 1180°C, then increases significantly after conversion to α-phase as grain growth occurs. The decrease provides advantageous cooling from 0.08Au to just inside 0.03Au. IR emittance decreases due to both physical coating changes and increasing temperature. At some higher temperature, pre-mission heat treated and in-situ heat treated coatings perform similarly in terms of heat shield temperature reduction.

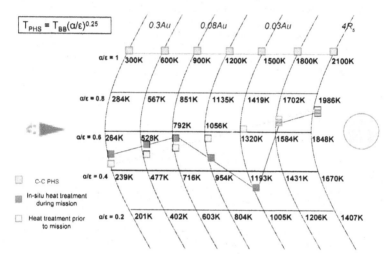

Figure 18. Solar Probe heat shield temperature during solar approach with different coating heat treatment scenarios

CONCLUSIONS

Heat treatments of Al_2O_3 coatings at critical phase transformation temperatures were performed to study microstructural changes and how these changes affect the optical properties of the coatings. The plasma-sprayed alumina coatings were deposited primarily as γ-phase material. Heat treatments at 900°C and above resulted in phase transformation from metastable γ-phase to stable α-phase, through intermediate phases. After 2 hours at 900°C and 1050°C, transformations from γ-Al_2O_3 to δ-Al_2O_3 were evident. After 2 hours at 1180°C, the Al_2O_3 coating was almost entirely α-phase material. Also due to annealing at 1180°C, major restructuring occurs and grains begin to form. Above 1180°C, the coating material is phase stabilized, but significant grain growth occurred during heat treatment at 1380°C and 1550°C. Changes in the microstructure and morphology caused significant changes in the optical properties of the coatings. Solar absorptance decreased from 0.29 in the as-sprayed condition to 0.21 after 2 hours at 900°C or 1050°C, and to 0.13 after 2 hours at 1180°C. Heat treatment at higher temperatures caused an increase in $α_S$, from 0.13 to 0.16 after 2 hours at 1380°C and to 0.23 after 2 hours at 1550°C. Emissivity of the coated C-C decreased with increasing heat treatment temperature. Emissivity also decreases as the temperature of the material increases, due to shifting of blackbody peak energy to lower wavelengths.

These changes in optical properties with heat treatment have a significant effect on the Solar Probe mission. Heat shield temperature is ultimately determined by the ratio $α_S/ε_{IR}$ of the coated C-C structure. Therefore, significant changes in these values during heat treatment of as-sprayed Al_2O_3 must be well understood. Plasma-sprayed coatings can either be stabilized prior to the mission or allowed to change phase and microstructure in-situ during the mission. Each of these options has advantages that must be considered further. Because the Solar Probe mission is enabled by favorable optical properties (low $α_S$ and high $ε_{IR}$), changes in these properties must be more well understood. Further study would include understanding the following: 1) changes in optical properties over a well defined mission trajectory with and without heat treatment prior to the mission; 2) the effect of sample temperature on the measured optical properties; 3) the ability to change the plasma spray process or inhibit grain growth to obtain different optical properties; and 4) the effect of microstructural and morphological changes on other coating aspects such as structural durability.

REFERENCES

1. Potocki, K., et al., "Solar Probe Engineering Concept," Solar Wind 11 – SOHO Conference, June 2005.
2. http://solarprobe.gsfc.nasa.gov/, *Solar Probe STDT Final Report*, prepared for NASA Headquarters, September 2005.
3. King, D.E., et al., "Al2O3 Optical Surface Heat Shield for Use in Near Solar Environment," *Proceedings of the 30th International Conference on Advanced Ceramics and Composites*, Cocoa Beach Fl, 2006.
4. Stumpf, H.C., et al., "Thermal Transformations of Aluminas and Alumina Hydrates," *Ind. Eng. Chem.*, **42**, 1398-1403, 1950.
5. Petersson, A., et al., "Microstructural Evolution and Creep Properties of Plasma Sprayed Nanocomposite Zirconia-Alumina Materials," *Proceedings of the 107th Annual Meeting of the American Ceramic Society*, 177, 61-71, 2006.
6. McPherson, R., "The Relationship Between The Mechanism of Formation, Microstructure and Properties of Plasma-Sprayed Coatings," *Thin Solid Films*, **83**, 297-310, 1981.
7. Damani, R.J. and A. Wanner, "Microstructure and elastic properties of plasma-sprayed alumina," *J. Materials Science*, **35**, 4307-4318, 2000.
8. Weferes, K. and C. Misra, "Oxides and Hydroxides of Aluminum," *Alcoa Technical Paper No. 19*, Alcoa Laboratories, 1987.
9. ASTM Standard E903-82, Standard Test Method for Solar Absorptance, Reflectance, and Transmittance of Materials Using Integrating Spheres. American Society for Testing and Materials, West Conshohoken, PA.

POROUS CERAMIC FOAM CATALYSTS FOR N_2O-BASED SATELLITE MICROTHRUSTERS

F. Ahmed, L. Courtney, J.R. Wallbank, P.A. Sermon

Chemistry, SBMS,
University of Surrey (UniS)
Guildford
Surrey, GU2 7XH, UK

ABSTRACT

A sol-gel Rh/TiO_2 coated onto an alumina ceramic foam has been shown to be stable and yet active in N_2O decomposition under conditions relevant to a microthruster for a small satellite. $/Al_2O_3$ (ceramic foam:CF). Its turnover frequency in N_2O decomposition was higher than for traditional catalysts. Routes to such porous ceramic catalysts by extrusion of alumina green bodies containing starch or sucrose are now being explored.

INTRODUCTION

Microthrusters in microspacecraft are required as micropropulsion systems for high-accuracy station keeping, attitude control, drag compensation and orbit adjust[1]. With time the dimensions, complexity, cost and mass (W) of satellites[2] decrease from mini- ($100kg<W<400kg$) to micro- ($10kg<W<100kg$) to nano-satellites ($1kg<W<10kg$). Yet all need to be able to change or maintain orbits and to compensate for atmospheric drag. Microthrusters can be based on solid propellants [3], (i) hot/cold gases[2] (e.g. $N_{2(g)}$ at 20MPa)[2], (ii) a coupled thermal-fluid[4], (iii) vaporizing or low boiling point liquids[5], (iv) decomposing N_2H_4, H_2O_2 or N_2O monopropellants in catalytic microthrusters[6], (v) bipropellants or (vi) plasmas[7].

Decomposition of monopropellant N_2H_4 (i.e. $N_2H_4 \rightarrow N_2 + 2H_2$) occurs over Shell 405[TM] catalyst (i.e. 29.7%Ir/Al_2O_3, $115m^2/g$, $PV= 0.18cm^3/g$, Ir area 33 m^2/g and spheres 0.6-0.8mm) at only 293K[8] (although the Shell 405 temperature then rises to 748K) causing an increase in fluid volume and thrust for essentially zero power initially supplied. H_2O_2 is said to be a greener monopropellant. It decomposes at lower temperatures on catalysts with no prior heating, but like N_2H_4 has safety issues. Low toxicity and high theoretical I_{sp} N_2O (i.e. $I_{SP(t)}$~206s) has also been suggested as an even greener monopropellant[9], but its decomposition is initiated at moderate temperature only. It can be stored at $\rho=0.75g/cm^3$ at 5MPa. UoSAT-12, launched at 6am on 21st April 1999, containing 2.5kg of N_2O for 14 h of resistojet propulsion[10]. An N_2O decomposition catalyst in this application needs to have activity at low temperature and yet thermal stability at high temperature and selectivity to product N_2 (since anthropogenic N_2O-derived NO[11] would help destroy stratospheric O_3). N_2O is thermodynamically unstable (e.g. for its dissociation lnK $=+42.0$ at 298K and $+18.8$ at 1000K), but is kinetically stable, since its bimolecular decomposition only takes place >838K with an activation energy of 245-264kJ/mol[12], because the conversion is spin-forbidden. N_2O decomposition is exothermic (i.e. $\Delta H=-81.55kJ/mol$ at 300K and $-82.38kJ/mol$ at 900K). It is catalysed by a variety of surfaces[13]. The kinetic orders with respect to N_2O and O_2 have been found to be +1.00 and -0.5, although retardation of the rate-determining step (*i.e.* O_2 release from the surface) by $O_{2(g)}$ depends on catalyst pre-treatment. Metals and oxides[13] are active in N_2O decomposition at temperatures far below those of the homogeneous bimolecular reaction. Since activation energies are lower on Rh_2O_3 (42kJ/mol) than Al_2O_3 (188kJ/mol), it is not surprising that oxide supports[13, 14] have their N_2O activity increased when supporting Rh[13]. N_2O has been used for propulsion: (i) in the 1930s both the British and German military experimented with N_2O as an oxidiser of solid/liquid fuels in rockets, (ii) in the 1950s the National Advisory Committee for Aeronautics commissioned a study of its use as a monopropellant, (iii) in the 1980s the AMROC corporation experimented with N_2O as an oxidiser in rocket engines and (iv) SpaceDev now uses N_2O as the oxidiser for their X-prize entry vehicle "Spaceship 1". Work on N_2O-based microthrusters has recently been taking place at UniS, Poitiers (France), St.Petersburg (Russia) and Tsinghua University (PRC). The development of an N_2O microthruster with a specific impulse (Isp) of ≈150s could involve an initial power requirement <31W.

Better catalysts are being sought (*e.g.* alternatives to Ir catalysts have been sought in nitrides and carbides of molybdenum and tungsten[15] and macroporous niobium oxynitride[8]) for new less toxic monopropellants[8] that will undergo exothermic decompositions in satellite microthrusters. Advances in materials and catalyst technology are necessary to nanoengineer these thrusters for smaller satellites. The authors believe that a nanoengineered ceramic catalyst may be of value. Specifically the authors are developing catalytic N_2O microthruster nanotechnology for satellites, where the catalysts are active in N_2O decomposition at low temperature (to lower power requirements) and yet stable at the high temperatures produced in oxidising conditions. Ceramics[16] are used as catalysts, catalyst supports, cellular foams (CF)[17], O^{2-} membranes[18], monoliths to promote partial oxidation of alkanes[19] and selective reduction of NO[20].

The authors wanted to determine if there was potential for low temperature ceramics in microthruster design [1], especially smaller MEMS-based microthrusters[21].

EXPERIMENTAL

Catalysts. Commercial 36%Ir/Al_2O_3 (Shell 405; CRI Fine Chemicals[22]) was used. This was compared with 0.05wt % Rh/TiO_2 (sol-gel-80 %)/Al_2O_3 (ceramic foam:CF). For this, white Al_2O_3 ceramic foam (CF; Hi-Por alumina provided by Dytech Corporation Ltd) was acid etched (3M HNO_3, 200 cm^3) three times for 15 h. A mixture of TiO_2 (sol-gel, 16 cm^3) and ethanol (0.087 mol, Hayman 99.99 %) was surface coated on CF block (1 cm x 1 cm x 3 cm) and dried for 16h. $RhCl_3.H_2O$ (0.118 mmol, Aldrich 99.98 %) was dissolved in ethanol (20 cm^3, 0.434 mol, Hayman 99.99 %) and added to the TiO_2 (SG-80 %)/Al_2O_3 CF and dried at 313-343 K. The product was a pink-red coloured ceramic block was then further dried at 373 K for 24 h.

Methods. X-ray photoelectron spectroscopy (XPS; VG Scientific) and scanning electron microscopy (SEM; Hitachi S-3200N) were used to characterise samples. The decomposition of N_2O was measured in terms of (i) temperature-programmed calorimetry or mass spectrometry of O_2 release and (ii) isothermal (523 or 973K) pulse decomposition. (i) involving a Setaram 121 differential scanning calorimeter (DSC) and an ESS VG quadrupole Sensorlab residual gas analyser (RGA) in combination. The catalyst was placed in the DSC, while 100cm^3/min N_2O + 50cm^3/min Ar flowed over it and then to the RGA as the sample temperature was raised at a controlled rate (10K/min) to 773K. In some experiments the RGA alone was used for samples (40-70mg pre-reduced in H_2 at 523K) of catalysts in a silica flow-microreactor. A reactant stream consisting of Ar (80 cm^{3}min^{-1}) + N_2O (20 cm^{3}min^{-1}) was introduced to the catalyst as this was heated in a Lenton PyroTherm furnace at 10K min^{-1} to 1073 K. The partial pressures of the following reactants and products were followed as a function of time and temperature at m/e = 32 (O_2), 31 (N_2O), 44 (N_2O), 28 (N_2) and 40 (Ar). In (ii) Ar flowed (80 cm^{3}min^{-1}) as duplicate 0.1cm^3 (2.24 mmol) N_2O pulses were introduced at 523 K and 973 K onto H_2 pre-reduced (673K; 50 ±1 cm^{3}min^{-1}) catalyst samples (0.05g). The number of moles of N_2O per mole of Rh_s in Rh/TiO_2 in each N_2O pulse (n_{N2O}/n_{Rhs}) was 0.22. In addition microthruster field tests were carried out on a I_{sp}=150 s N_2O prototype thruster. This was designed to simulate a lightweight 100 mN thruster of a microsatellite. The pressure (0.3MPa) and flow (4 cm^{3}min^{-1}) rate of the N_2O, the power (30 W) to the heater and the temperature (300-1400K) within the catalyst bed were all measured as a function of time (0<t<1000s).

CHARACTERISATION RESULTS

Figure 1 suggests that Ir is well distributed on the surface of the commercial alumina in the Shell 405 catalyst before it is used in N_2O decomposition, but that this is not the case after N_2O decomposition catalysis. XPS found that 48 % of original Ir was lost on N_2O reaction (See Table 1). Clearly the reaction causes some sintering and loss of Ir on the alumina support as a result of the N_2O reaction. Figure 2 gives characterization data for the sol-gel coated alumina ceramic foam that contained just 0.5%Rh. Clearly the porous alumina ceramic foam has been coated with the sol-gel catalyst. The metal losses and oxidation states in Shell 405 and Rh/TiO_2/Al_2O_3 (CF) are shown in Table 1; both Ir and Rh were transformed to positive oxidation states by the reaction, but more Ir than Rh lost from the catalyst.

Table 1. XPS-derived metal loss and oxidation state			
Catalyst	% metal before reaction	% metal after reaction	Oxidation state change
Rh/TiO$_2$/Al$_2$O$_3$ (CF)	0.23	0.20	+3 $\Rightarrow$ +4
Ir/Al$_2$O$_3$ – Shell 405	4.74	2.28	+1 $\Rightarrow$ +4

a

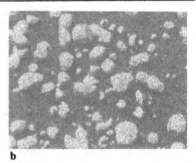

b

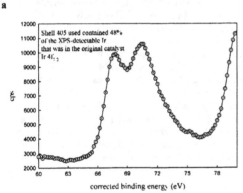

c

Figure 1. EDX maps of Ir across a 30μm x 20μm area of Shell 405, a: unused) and b: after N$_2$O decomposition. c: X-ray photoelectron spectroscopy (XPS) suggested that Ir^{4+} was present in Shell 405 after use, but that 48% of the Ir *was lost* from the catalyst on use in N$_2$O decomposition.

CATALYTIC RESULTS

Figure 3a shows that Shell 405^{TM} is active in N$_2$O decomposition at 560K with N$_2$O consumption, N$_2$ liberation, O$_2$ liberation and heat liberation. There is reversibility to temperature change in that no activity-temperature hysteresis is exhibited on thermal cycling. From the rates of heat release seen over Shell 405 by DSC (and rates of heat release deduced from RGA analysis of [N$_2$O]) plotted in a pseudo-Arrhenius manner the activation energies were found to be:

DSC E$_a$=160.6 kJ mol^{-1}
RGA E$_a$=115.3 kJ mol^{-1}

and so diffusion limitation is thought to be absent when judged by both techniques. The results also mean that calorimetry and residual gas analysis are measuring different rates of reaction and this that methods of analysis in scanning for high-throughput methods of optimising catalysts need to be carefully chosen.

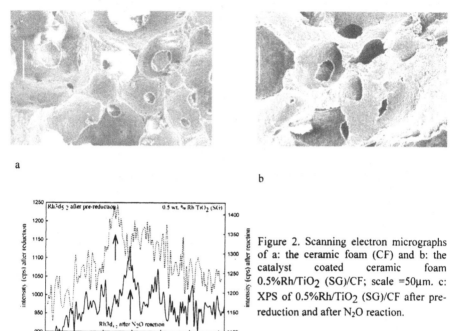

a

b

c

Figure 2. Scanning electron micrographs of a: the ceramic foam (CF) and b: the catalyst coated ceramic foam 0.5%Rh/TiO$_2$ (SG)/CF; scale =50μm. c: XPS of 0.5%Rh/TiO$_2$ (SG)/CF after pre-reduction and after N$_2$O reaction.

Figure 3b also shows that the Rh/TiO$_2$ sol-gel coated alumina ceramic foam is active but at a higher temperature. Table 2 compares the Shell 405 and ceramic catalysts, in terms of activation energies and turn over frequencies (TOFs). Clearly the ceramic foam catalyst has a lower activity per unit mass and a higher activation energy than Shell 405. However, Table 2 also shows that it does have a higher turnover frequency (TOF) in N$_2$O decomposition, and this catalyst is also more stable at these temperatures under such oxidizing conditions (see Figure 2).

Table 2. Activation energies and turnover frequencies		
Catalyst	E_a (kJ/mol)	TOF at 900K (s^{-1})
Shell 405	161	0.15
0.5wt% Rh/TiO₂/ Al₂O₃ ceramic foam	217	9.27

* the apparent turnover number was measured as the number of O_2 molecules produced from N_2O at 900K per second per metal atom

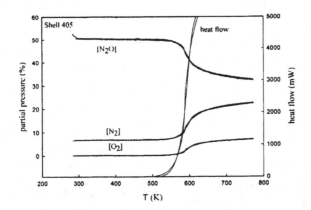

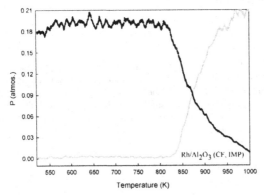

Figure 3. N₂O decomposition activity of Shell 405 and Rh/TiO₂/Al₂O₃ (CF)

While in terms of activity at low temperatures in the first catalytic run one has Shell 405 > Rh/TiO₂/Al₂O₃ (CF), the ceramic foam catalyst is superior in terms of rates per active metal ion then the order is reversed: Shell 405 < Rh/TiO₂/Al₂O₃ (CF).

Pulse N_2O isothermal experiments (see Figure 4) confirm that N_2O decomposes on the coating of the Rh/TiO₂/Al₂O₃ (CF) at 973K but not 573K (Figure 3). This is relevant to intermittent microthruster use. Such a temperature needs lowering. Interestingly, product O_2 is not released in these short timescales, but is retained by the catalyst as Rh is oxidized (+1 → +3).

573K

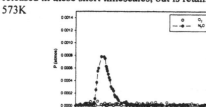

973K

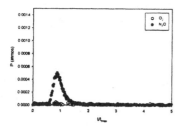

Figure 4. Decomposition of coating on Rh/TiO₂/Al₂O₃ (CF). N_2O pulses onto 0.5wt % Rh/TiO₂ (SG) in flowing Ar at 523K (left) and 973K (right). Reaction occurs at 973K decomposing N_2O, but is not liberating O_2 (Rh=>RhO$_x$) at low reaction times.

DISCUSSION AND CONCLUSIONS

The authors have shown that while the Shell 405 is good in the short term in N_2O microthrusters, Ir losses are significant and that better turnover frequencies and lower metal losses are seen for ceramic foam catalysts (e.g. Rh/TiO₂/ceramic foam (CF)). The authors believe these can now be developed for N_2O microthrusters for space applications. The authors want to explore production of such catalysts from ceramic green bodies which contain organic modifiers that define the porosity and pore sizes. The authors have considered PMMA microspheres with oxide overcoats[23], and extrudates from chilled sepiolite-alumina-starch-sucrose-water (60wt%)[24] green bodies and then novel porous ceramic catalyst monoliths for the microthrusters (see Figure 5), on calcination at 373K and 1673K. Initial experiments are promising and will be reported.

a   b

Figure 5. (a) Extrudate of chilled alumina catalyst-sepiolite-starch-water (60wt%) are being evaluated. (b) porous ceramic from alumina and sucrose.[24]

ACKNOWLEDGEMENTS

The authors acknowledge support from EPSRC and Dytech International for JW, the support of Mechadyne International and KTP for LC and the provision of the extrudates of UniS catalysts by Malcolm Yates (UAM, Madrid). They also acknowledge useful discussions with D.Gibbon (Surrey Satellite Technology Ltd).

REFERENCES
[1]K.L.Zhang, S.K.Chou and S.S.Ang *J.Micromechanics and Microengineering* **15**,944,(2005)
[2]J.Kohler, J.Bejhed, H.Kratz, F.Bruhn, U.Lindberg, K.Hjort and L.Stenmark *Sens. Actuators* **97-8A**,587-598,(2002); K.L.Williams, A.B.Eriksson, R.Thorslund, J.Kohler, M.Boman and L.Stenmark *J.Micromechanics and Microengineering* **16**,1154,(2006)
[3]C.Rossi, D.Briand, M.Dumonteuil, T.Camps. P.Q.Pham and N.F.de Rooij *Sensors and Actuators* **126A**,241,(2006)
[4]A.A.Alexeenko, D.A.Levin, A.A.Fedosov, S.F.Gimelshein and R.J.Collins *J.Propulsion and Power* **21**,95,(2005)
[5]E.V.Mukerjee, A.P.Wallace, K.Y.Yan, D.W.Howard, R.L.Smith and S.D.Collins, *Sen. Actuators* **83A**,231-236,(2000); T.G.Kang, S.W.Kim and Y.H.Cho *Sens. Actuators* **97-8A**,659-664,(2002); D.K.Maurya, S.Das and S.K.Lahiri *Sens. Actuators* **122A**,159,(2005)
[6]H.E.Barber, G.L.Falkenst, C.A.Buell and R.N.Gurnitz *J.Spacecraft and Rockets* **8**,111,(1971)
[7]C.Phipps, J.Luke and T.Lippert Thin Solid Films 453-454,573,(2004); A.Kakami, H.Koizumi, K.Komurasaki and Y.Arakawa *Vacuum* 73,419,(2004)
[8]R.Brayner, G.Djega-Mariadassou, G.M.daCruz and J.A.J.Rodrigues *Catal.Today*, **57**,225,(2000)
[9]T.Lawrence and M.Caporicci *Proc. Intern.Conf. Green Prop.* (ESA/ESTEC, 2001); Zakirov *et al.. Acta Astronautica*, **48**, p353, 2000; M.N.Sweeting, T.Lawrence and J.Leduc *Proc. Inst. Mech. Eng.* **213C(G4)**,223,(1999); P.A.Sermon, D.M.Gibbon, C.Euesden,V.Zakirov, C.Huggins, M.A.M.Luengo and R.Sambrook (Holland 2001); Zakirov, PhD Thesis, University of Surrey, UK, 2001.
[10]C.W.Bostian, W.T.Brandon. A.U.MacRae, C.E.Mahle and S.A.Townes *Space Comm.* **16**.97,(2000)

[11]W.C.Trogler *Coordn.Chem.Rev.* **187**,303,(1999); K.S.Bradley, K.B.Brooks, L.K.Hubbard, P.J.Popp and D.H.Sedman *Env.Sci.Tech.* **34**,897,(2000)

[12]C.N.Hinshelwood and R.E.Burk *Proc.Roy.Soc.* **106A**,284,(1924)

[13]G.Centi, L.Dallolio and S.Perathoner *Appld.Catal.* **194A**,SISI,79,(2000); S.Kameoka, T.Suzuki, K.Yuzaki, T.Takeda, S.Tanaka, S.Ito, T.Miyadera and K.Kunimori *J.Chem.Soc.Chem.Commun.* 745,(2000); E.R.S.Winter, *J.Cat.* **19**,41,(1970);R.Larsson *Cat.Today* **4**,235,(1989)

[14]T.M.Miller and V.H.Grassian *J.Amer.Chem.Soc.* **117**,10969,(1995)

[15]J.A.J.Rodrigues, G.M.Cruz, G.Bugli, M.Boudart and G.DjegaMariadassou *Catal.Lett.* **45**,1,(1997)

[16]P.Kolsch, M.Noack, R.Schafer, G.Georgi, R.Omorjan and J.Caro *J.Membrane Sci.* **198**,119,(2002); M.J.Ledoux and C.P.Huu *Cattech* **5**,226,(2001)

[17]Y.A.Aleksandrov, I.A.Vorozheikin, K.E.Ivanovskaya and E.I.Tsyanova *Russ.J.Gen.Chem.* **71**,825,(2001); C. van Gulijk, M.Makkee and J.A.Moulijn *Topics Catal.* **16**,285,(2001); F.C.Buciuman and B.K.Czarnetzki *Catal. Today* **69**,337,(2001)

[18]Z.P.Shao, G.X.Xiong, H.Dong, W.H.Yang and L.W.Lin *Sepn.Purifn.Technol.* **25**,97,(2001)

[19]D.Farrusseng, A.Julbe and C.Guizard *Sepn.Purifn.Technol.* **25**,137,(2001); L.J.Wang, S.H.Ge, C.H.Liu and Z.H.Li *J.Porous Media* **4**,253,(2001)

[20]T.Valdes-Solis, G.Marban and A.B.Fuertes *Catal.Today* **69**,259,(2001)

[21]K.L.Zhang, S.K.Chou and S.S.Ang *J.Microelectromechanical Systems* **13**,165,(2004); C.Rossi, T.Do Conto, D.Esteve and B.Larangot *Smart Materials & Structures* **10**,1156,(2001); C.Rossi, D.Esteve and C.Mingues *Sens.Actuators* **74A**,211,(1999)

[22]Shell.Tech.Bull. (August 1988)

[23]F.Tang *J.Euro.Ceram.Soc.* **24**,341,(2004); D.J.Wang *Chinese Chem.Lett.* **14**,1306,(2003); D.J.Wang *Colloid Poly.Sci.* **282**,48,(2003); D.J.Wang *Chem.Lett.* **32**,36,(2003)

[24]K.Prabhakaran, N.Madhusudan Gokhale, S.C.Sharma and R.Lal *J.Amer.Ceramic Soc.* **88**,2600,(2005)

Multifunctional Coatings, Nanostructured Coatings, and Interfaces Phenomena

DEVELOPMENT OF MULTI-LAYERED EBC FOR SILICON NITRIDE CERAMICS

Shunkichi UENO[1], Tatsuki OHJI[2], and Hua -Tay LIN[3]

[1]The Institute of Scientific and Industrial research, Osaka University, 8-1 Mihogaoka, Ibaraki, Osaka 567-0047, Japan
[2]Advanced Manufacturing Research Institute, National Institute of Advanced Industrial Science and Technology, 2268-1 Shimo-Shidami, Moriyama-ku, Nagoya 463-8687, Japan
[3]Materials Science and Technology Division, Oak Ridge National Laboratory, Oak Ridge, TN 37831-6068, USA

ABSTRACT
In the development of ceramic components for land-based micro gas turbines, we designed a silicon nitride ceramic with multi-layered environmental barrier coating (EBC) system that was based on the results of corrosion and/or recession tests for each component material. In this paper, results of corrosion and/or recession tests for each material, preparation of silicon nitride with multi-layered EBC, and recession test for the silicon nitride with EBC were discussed. The multi-layered EBC system was constructed with thin $Lu_2Si_2O_7$ as first layer, $Lu_2Si_2O_7$/mullite eutectic as an intermediate layer, and thick $Lu_2Si_2O_7$ as a top-coat layer. From the static state water vapor corrosion tests for oxides with low coefficient of thermal expansion at $1300^\circ C$, the corrosion resistances for $Lu_2Si_2O_7$ was far superior to rest of the oxides evaluated. However, a selective corrosion of boundary phase in $Lu_2Si_2O_7$ was observed. To eliminate this problem, a $Lu_2Si_2O_7$/mullite eutectic without silica-containing boundary phase was developed. The corrosion rate of this bulk was not measurable within experimental error. On the other hand, the recession tests for these EBC samples were performed at $1300^\circ C$ in high-speed steam jet. For $Lu_2Si_2O_7$, the phase decomposed into Lu_2SiO_5 and a porous structure was formed on the bulk surface. For $Lu_2Si_2O_7$/Mullite eutectic bulk, mullite phase was completely removed from the bulk surface. Hence, the $Lu_2Si_2O_7$/mullite eutectic was then employed as an intermediate layer in the multi-layered EBC system aiming to effectively inhibit the inward diffusion of water vapor. Results showed that the multi-layered EBC system was well sustained in high-speed steam jet environment at $1300^\circ C$ for 500 hr.

INTRODUCTION
On the development of Si-based ceramic components for land-based micro gas turbines, improvement of EBC system, which protects the non-oxide ceramics substrate from oxidation and/or water vapor corrosion, has become a main issue in the development. Several research groups proposed multi-layered EBC system for non-oxide ceramics [1,2]. Since the coefficient of thermal expansion (CTE) of non-oxide ceramics such as silicon nitride, silicon carbide and their composites are relative low, the CTE of first layer of the multi-layered EBC system must be comparable to that of the substrate. In some of the previous studies, silicon [1] and zircon [2] are used as the first layer because of their low CTE values. In Table 1, several low CTE materials are summarized. In the multi-layered EBC system, it's anticipated that the first layer would not encounter a direct attack of high-speed water vapor. Hence, corrosion data of materials in static state water vapor environment could be useful inputs for designing the first or intermediate EBC layer. As results the authors have examined static state water vapor corrosion resistances for low

CTE materials [3-5]. On the other hand, the top layer of multi-layered EBC system would expose to high velocity water vapor in combustion environment. Hence, the material using as the top coat must sustain the long-term exposure to high velocity combustion gas for at least for one year in land-based micro gas turbines. Thus, recession tests for some oxides under high speed steam jet environment were also carried out by the authors.

In this paper, a new concept of multi-layered EBC system for silicon nitride ceramics is proposed and the results of a recession test for silicon nitride ceramics with the multi-layered EBC are discussed.

EXPERIMENTAL PROCEDURES

The SN-282 silicon nitride ceramics (manufactured by Kyocera Co. Ltd, Kagoshima, Japan) containing lutetium oxide (Lu_2O_3) as sintering additive was used as the substrate. The multi-layer EBC consisted of a thin $Lu_2Si_2O_7$ layer bond coat, thick $Lu_2Si_2O_7$/mullite eutectic intermediate layer, and a thick $Lu_2Si_2O_7$ top layer. The first layer was coated by a new oxidation bonded by reaction sintering (OBRS) method on the silicon nitride substrate, which was developed by the authors [3].

The deposition of the $Lu_2Si_2O_7$/mullite eutectic layer and $Lu_2Si_2O_7$ top coat layer was carried out using the air plasma spray (APS) method. The mixed powders of high purity Lu_2O_3 (99.99% purity, Shin-Etsu Chemical Co. Ltd., 4μm particle size), SiO_2 (99.99% purity, High Purity Chemicals Co. Ltd., 0.8μm particle size), and Al_2O_3 (99.99% purity, High Purity Chemicals Co. Ltd., 0.8μm particle size) were used for APS coating. For the preparation of eutectic layer, the powder mixture with the composition of Al_2O_3:SiO_2:Lu_2O_3 = 27.3:54.6:18.1 in molar ratios [6] was used. For the preparation of $Lu_2Si_2O_7$ top coat layer, the powder mixture with the composition of SiO_2:Lu_2O_3 = 2.0:1.0 in stoichiometric molar ratios was used.

The recession test of the prepared sample was performed at 1300°C for 500 hours using a water injection facility at Oak Ridge National Laboratory (ORNL). The distill water was heated up to ~ 200°C and directly sprayed on the sample surface via a water pump. The velocity of the steam jet was estimated to be ~50 m/s [7].

RESULTS AND DISCUSSION

Table 2 summarized the water vapor corrosion rates for low CTE materials that were obtained by the corrosion test in static state water vapor environments by the authors. Based on the static state water vapor exposure results, the corrosion resistance of zircon phase is excellent. However, the CTE for zircon phase is slightly larger than that of silicon nitride ceramics, which could lead to thermal mismatch issue after deposition. Thus, the $Lu_2Si_2O_7$ phase would be a more compatible system the first layer in multi-layered EBC system for silicon nitride ceramics.

A high density $Lu_2Si_2O_7$ layer with ~5 μm thickness was successfully prepared on silicon nitride substrate via the OBRS method. In this method, silicon nitride ceramics were heat-treated at relatively low temperatures, i.e. at below 1100°C, and a thin amorphous silica layer (~ 5 μm) formed on the surface due to the oxidation. For coating of $Lu_2Si_2O_7$ layer on silicon nitride ceramics, the silicon nitride substrate with a thin layer of SiO_2 was packed with the Lu_2O_3 powder bed and sintered at 1600°C for 4 hrs in argon atmosphere (0.5 MPa). A thin $Lu_2Si_2O_7$ layer without cracking can then be prepared by this method. The detailed procedures of the preparation of the thin $Lu_2Si_2O_7$ layer could be found in the previously published paper [3].

The X-ray diffraction pattern of the as-deposited APS top coat layer showed Lu_2O_3 and amorphous phases [8]. The center position of the broad peak indicated the amorphous phase was

$2\theta=30°$. Commonly, a broad peak for amorphous silica appears around $2\theta=15\text{-}30°$. Hence, it was suggested that the composition of the amorphous phase contained Lu_2O_3 component [8]. Consequently, a post heat treatment was performed at $1500°C$ in argon for 4 hours to crystallize the top coat. By this heat treatment, Lu_2SiO_5 and $Lu_2Si_2O_7$ mixed crystalline layer was obtained.

The multi-layer EBC did not spall during the steam jet test. Figs.1 (a) and (b) show SEM micrographs of surface and transverse cross section views of the sample after the steam jet test. On the surface, a porous structure associated with many cracks was formed. Results of previous recession test of $Lu_2Si_2O_7$ bulk via the steam jet test showed that a porous structure formed on the bulk surface due to the corrosion of boundary phase and silica component from $Lu_2Si_2O_7$ according to equation (1) [9].

$$Lu_2Si_2O_7 + 2H_2O = Lu_2SiO_5 + Si(OH)_4 \qquad (1)$$

In this case, the porous structure formed on the top coat could be attributed to the same corrosion mechanism. From cross section view, it can be confirmed that the cracks induced on the coating surface propagated through the layers and reached the substrate. Thus, a thin layer of thermally grown oxide was formed, as shown in the figure, due to the oxidation of substrate.

From this recession test, it was confirmed that even though a porous micro structure was formed on the top coat, the multi-layered EBC system, in general, was well sustained during the high speed steam jet exposure at $1300°C$. Nonetheless, many cracks through the coating system were generated during the test that led to the oxidation of the substrate owing to the inward diffusion of oxygen and/or water vapor. Thus, more research efforts to further optimize the multi-layered EBC systems need to be carried out to ensure long-term microstructure and chemical stability of coating components and thus lifetime performance of silicon nitride components under combustion environments.

CONCLUSIONS

The silicon nitride ceramics with multi-layered EBC system well sustained under high velocity steam jet environment at $1300°C$ for 500 h. The surface of the top coat was corroded by water vapor, consistent with the recession test of the bulks. However, the multi-layered EBC system well protected the corrosion and/or oxidation of silicon nitride substrate. Some cracks that pass through the layers were induced during the test.

REFERENCES

[1]K.N. Lee, D.S. Fox, J.I. Eldridge, D. Zhu, R.C. Robinson, N.P. Bansal, R.A. Miller, "Upper Temperature Limit of Environmental Barrier Coatings Based on Mullite and BSAS," *J. Am. Ceram. Soc.*, **86**, 1299-1306 (2003).

[2]H. Kose, Y. Yoshiaki, H. Kobayashi, T. Takahashi, Kokaitokkyokoho, JP5-238859.

[3]S. Ueno, D.D. Jayaseelan, T. Ohji, "Development of Oxide-Based EBC for Silicon Nitride," *Int. J. Appl. Ceram. Technol.*, **1**, 362-373(2004).

[4]S. Ueno, T. Ohji, H.T. Lin, "Corrosion and recession behavior of zircon in water vapor environment at high temperature," *Corrosion Science*, **49**, 1162-1171 (2007).

[6]S. Ueno, D.D. Jayaseelan, T. Ohji and H.T. Lin, "Recession behavior of $Lu_2Si_2O_7$/mullite eutectic in steam jet at high temperature," *J. Mater. Sci.*, **40**, 2643-2644 (2005).

[7]H.T. Lin and M.K. Ferber, "Mechanical reliability evaluation of silicon nitride ceramic components after exposure in industrial gas turbines," *J. Europ. Ceram. Soc.*, **22**, 2789-2797 (2002).

[8]S. Ueno, T. Ohji and H.T. Lin, "Preparation and recession test of multi-layered EBC system," *Ceram. Int.*, in press.

[9]S. Ueno, D.D. Jayaseelan, T. Ohji and H.T. Lin, "Recession mechanism of $Lu_2Si_2O_7$ phase in high speed steam jet environment at high temperatures," *Ceram. Int.*, **32**, 775-778 (2006).

Table 1   Candidate EBC materials and their CTE

	CTE ($/°C \times 10^{-6}$)
Al_2TiO_5	0
mullite	4.5-
hafnon	3.6
zircon	4.43
$Yb_2Si_2O_7$	3.62
$Lu_2Si_2O_7$	3.84
(substrate,SN-282	3.54)

Table 2 Corrosion rate for EBC materials

	CTE ($/°C \times 10^{-6}$)	weight loss rate ($g/cm^2 h \times 10^{-6}$)
hafnon	3.6	7.1
zircon	4.43	1.3
$Yb_2Si_2O_7$	3.62	7.5
$Lu_2Si_2O_7$	3.84	4.2

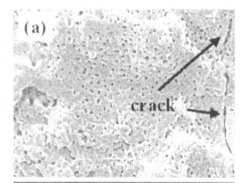

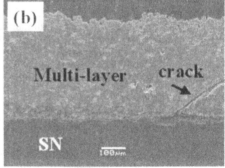

Fig.1 The sample after the test.

REACTIVE BONDING OF SAPPHIRE SINGLE CRYSTAL TO TUNGSTEN-COPPER METAL COMPOSITE USING DIRECTED VAPOR DEPOSITION PROCESS

Y. T. Peng[1], D. D. Hass[2], and Y.V. Murty[1]

[1]Cellular Materials International, Inc.
2 Boar's Head Lane
Charlottesville, VA 22903, USA

[2]Directed Vapor Technologies International, Inc.
2 Boar's Head Lane
Charlottesville, VA 22903, USA

ABSTRACT

A method of reactively bonding the metal-ceramic interfaces using a modified electron-beam physical vapor deposition process called Directed Vapor Deposition (DVD) developed by Directed Vapor Technologies International (DVTI) is described in this paper. This process has been effectively employed by depositing Ti and Ag layers in the selected areas of a complex shape sapphire substrate to promote robust reaction bonded interface between sapphire and a coefficient of thermal expansion (CTE) matching W/Cu composite. At locations on the interface where the metallized sapphire is adjacent to the Cu matrix of W/Cu composite, Energy Dispersive X-ray Spectroscopy (EDS) results indicate an extensive degree of Ag diffusion into the Cu matrix, as well as the reverse diffusion of Cu into the Ag coating, resulting in an indistinguishable bonding interface. At locations where W particles from the composite are in contact with the metallized sapphire substrate, there is no appreciable elemental inter-diffusion observed. Details of the DVD process, characterization of the reaction bonded interface, and its application to the ceramic/metal bonding was discussed in detail.

INTRODUCTION

Bonding/Brazing ceramics to metals presents a challenging opportunity due to vast variation of their physical properties such as high temperature atomic diffusivity, CTE mismatch, melting point variation, and their differences in reactivity at elevated temperatures in general. Significant interest exists to join metals to ceramics to produce cost effective engineered components for a number of electronic, automotive, chemical, and other industrial applications. Past work has shown that the joining of specific combinations of metal and ceramic systems can be achieved using mechanical and/or foil based brazing techniques. Such approaches, however, add complexity, cost, and weight. It also complicates the reliability issues as well. Conventional brazing processes require specialized brazing foils/paste with appropriate fluxes and binders. This gives raise to contamination and other performance degradation issues. Therefore, there is a constant interest in developing cost effective and reliable joining processes. In this paper a method of joining sapphire to metal is presented using a thin film reaction bonding approach with a modified electron-beam physical vapor deposition process, and thus eliminating conventional brazing.

Directed Vapor Deposition (DVD) is a plasma-assisted deposition technology capable of rapidly creating high quality coatings [1]. In this approach E-beam evaporated source material is entrained in a supersonic gas jet and focused towards the substrate, resulting in enhanced deposition rates and more efficient utilization of target material. Controlling the chamber pressure and pressure ratio of gas jet enables non-line-of-sight deposition where the sides and back of the substrate can be coated [2]. Conventionally, ceramic materials can be joined with metal in a two step metallization process where the ceramic surface is first coated with a layer of metal to improve wettability followed by a brazing step using filler metals. However, for parts with non-planar surfaces, uniform metal coating is difficult using methods such as sputtering or thermal evaporation without complicated substrate rotation or the use of multiple sources. The unique properties of DVD offer a novel opportunity in creating uniform and high-performance coatings on complex shape parts to aid in ceramic/metal joining.

The current study is focused on developing a bonding approach for a sapphire/metal electrode system in order to study plasma arc characteristics at high temperature in a very high gas flow field inside a confined geometry. Single crystal sapphire was selected because of its high temperature properties as well as its optical transparency characteristics on polished surfaces. For metal electrode a W/Cu composite was selected due to its excellent resistance to electric erosion compared to pure copper in high-current arc discharge applications [3]. Electrode made of copper, while having very high electrical and thermal conductivity, degrades rapidly due to evaporation or partial melting of copper on the electrode surface. With W/Cu composite high-melting-point W particles are embedded in a Cu matrix, creating a microstructure capable of providing simultaneous high conductivity and resistance to electric arc erosion. One major issue of metal-ceramic joining is the mismatch of thermal expansion. Residual stress developed at the bonding interface of mismatch materials can significantly reduce bonding strength [4]. The fact that the CTE of W/Cu composite can be adjusted by composition greatly facilitates its joining to sapphire.

To improve the strength of metal-ceramic bonding, the ceramic is first coated with an active metal such as Ti, which can serve as a "gluing" layer between ceramic and metal coating due to the capability of Ti in forming strong chemical bonds with the oxygen in ceramic at ambient temperature [5]. A thin second layer of more noble metal such as Ag is coated over Ti to protect the Ti from oxidation and serve as the substrate for subsequent bonding. In the current study, without this second Ag layer, titanium oxide would have prevented adhesion to metal on the electrode side. In fact, a type of Ag-Cu alloy containing approximately 2% Ti under the trade name Lucanex 721 is widely accepted as an active filler metal for brazing ceramic to metal [6]. However, because of this special formulation, the availability of this alloy is limited and expensive.

The process described herein, describes reactively bonding a pair of W/Cu electrodes to a complex shaped sapphire substrate using DVD technique. The electrode structure is made by sequentially depositing Ti and Ag thin films to the sapphire channel walls, whose surfaces are almost out-of-sight to the vapor source, followed by diffusion bonding Ag-coated W/Cu electrodes into the channels. The CTE of W/Cu electrode is adjusted to match that of the sapphire substrate to improve compatibility. The bonding interface between sapphire and W/Cu electrode is examined using Scanning Electron Microscopy (SEM) and Energy Dispersive X-ray Spectroscopy (EDS). Elemental distribution results from EDS indicate an extensive diffusion of Ag into Cu matrix of the composite, as well as the reverse diffusion of Cu into Ag coating, resulting in an indistinguishable bonding interface.

EXPERIMENTAL DESCRIPTION

Figure 1 shows a sapphire electrode assembly embedded and reaction bonded to a W/Cu electrode assembly to be used in the experiments to control the boundary layer separation during supersonic flow at Princeton University [7]. As mentioned before, sapphire was chosen as the substrate material for its high melting point, transparency, strength, and robustness in harsh environment. The sapphire preform was machined to the desired substrate configurations. Two channels of 13 mm long and 5 mm deep were machined into the sapphire substrate to accommodate the electrodes. DVD was used to create uniform coating on the bottom and walls of the sapphire channels. The CTE of W/Cu electrode was adjusted to about $8 \times 10^{-6}/°C$ with a composition of 80W-20Cu weight pct (34 volume pct of Cu) [8] to match sapphire substrate's CTE. The W/Cu composite was machined to a tight fit into the sapphire channels.

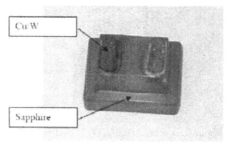

Figure 1. Sapphire electrode assembly reaction bonded with DVD process.

A schematic illustration of a DVD coating system is shown in Figure 2. Non-line-of-sight deposition is achieved by vapor scattering in the laminar flow surrounding the substrate at a relatively high chamber pressure [2]. Detail design features of a DVD system can be found elsewhere [9]. Ti vapor was generated by electron beam evaporation from a 12.7 mm diameter Ti source rod (five 9's) and deposited onto the sapphire substrate positioned about 20.0 cm away from the nozzle exit. The opening of sapphire channels was oriented to face down towards the center of the vapor source during deposition. Ag vapor was deposited in a similar fashion following the Ti deposition without breaking the vacuum. The sapphire substrate was kept at about 300°C during the deposition process. Coatings were produced with a pressure ratio of 5.8 with the gas jet pressure upstream the nozzle at 64 Pa and the chamber pressure at 11 Pa. The nozzle opening diameter was 30 mm in all experiments. Electron beam currents of 36 mA and 75 mA were used for Ti and Ag evaporation, respectively. The deposition time was 10 minutes for Ti and 20 minutes for Ag. After Ag coating the W/Cu composite was diffusion bonded within the sapphire channel at 850°C for 20 minutes under vacuum.

Samples polished using diamond lapping films with abrasives of sizes down to 0.5 μm was examined in a Jeol JSM 840 SEM fitted with an X-ray detector made by PGT Princeton Gamma Tech. The SEM images and EDS line scans were analyzed using PGT Spirit software. Nanoindentation tests were performed at the interfaces of W/Cu composite and sapphire using a Triboscope (Hysitron Inc.) in conjunction with a Veeco Dimension 3100 AFM system (Veeco Metrology Group).

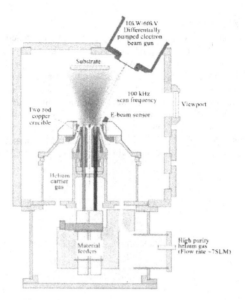

Figure 2. A schematic of a DVD coating system

RESULTS AND DISCUSSION

A typical cross sectional area of the bonding interface between W/Cu composite and sapphire from SEM measurement is shown in Figure 3. The W/Cu composite consists of W particles embedded in a continuous Cu matrix, while the sapphire at the top has a more uniform appearance. The surface of metallized sapphire substrate is either joined with the Cu matrix part of the composite or is in contact with the W particle part of the composite. EDS measurements were performed in these two types of interfaces to study the extent of elemental inter-diffusion. At certain portions of the interface (as seen at the right-hand side of the interface in Figure 3) there are gaps separating sapphire from W/Cu electrode, caused most likely by the rough and uneven finish of sapphire channel walls and the electrodes from the machining operation. Diffusion bonding could not be successfully accomplished due to insufficient intimate contact of surfaces in these areas.

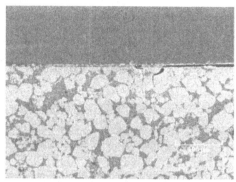

Figure 3. A cross-sectional area of the W/Cu composite and sapphire interface

SEM and an EDS line scan were performed in an area of the sapphire/electrode interface where the Ti/Ag coated sapphire surface is in direct contact with an elongated W particle. The SEM image of this area is shown in Figure 4 (a). The sapphire surface is coated with a Ti layer of about 2 μm thick (between point A and B in Figure 4 (a)) with a column like structure and a more porous layer of Ag coating (from point B to C in Figure 4 (a)) of about 6 μm thick. In contact with the Ag coating is an elongated W particle (from point C to D in Figure 4 (a)), where a clearly visible interface is present separating this elongated W particle and the Ag coating at point C. Below this elongated W particle is a thin section of Cu matrix (from point D to E in Figure 4 (a)) sandwiched by another W particle beneath (below the point E in Figure 4 (a)). The transition between this section of Cu matrix and its neighboring W particles appears to be gradual, and a clear-cut interface is absent. An EDS line scan of Ti-Kα, Cu-Kα, Ag-Lα, and W-Lα peaks was performed to confirm the above observations and to study the elemental inter-diffusion at the interfaces.

The EDS results are shown in Figure 4 (b). Major peak locations for all the elements are highlighted for ease of virtualization. The locations A, B, C, D, and E with associated dotted vertical lines in Figure 4 (b) correspond to the similarly labeled positions in Figure 4 (a). As can be seen in Figure 4 (b), the highlighted peak location for each element generally agrees with the SEM observation where a Ti/Ag coated sapphire piece is in contact with a W particle with a sandwiched Cu matrix underneath the W particle. At the boundary location "C" in Figure 4 (b) where the Ag coating is in contact with an elongated W particle, there is about 1 μm spatial overlapping between the Ag-Lα and W-Lα peaks. Since W and Ag are mutually immiscible [10], the degree of peak overlapping of about 1 μm can be interpreted as the EDS spatial resolution at this location. It is to be noted that this interface is clearly defined in the SEM image in Figure 4 (a). Due to the limited degree of Ag and W inter-diffusion, it is believed that strong sapphire-composite bonding is absent in this type of interfacial areas. It is interesting to note that at the boundary location "D" in Figure 4 (b) between the elongated W particle and Cu matrix, where a gradual transition between the two phases was observed in the SEM image in Figure 4 (a), there is relatively extensive W-Lα and Cu-Kα peak spatial overlapping extending for about 3 μm. Since W and Cu are also mutually immiscible [10], it is believed that this peak overlapping is caused by the distortion and/or spreading of Cu matrix and/or W particles, resulting in the intermixing of both phases. Certain steps in sample preparation such as forming, cutting, or polishing may have contributed to the distortion of either or both phases. Another interface of interest is the one between Ti and Ag coating at location "*B*" in Figure 4 (b), where the Ag-Lα peak spreads extensively into the Ti-Kα peak while the Ti-Kα peak also spreads to a certain degree into the Ag-Lα peak. The extent of spatial overlapping is about 4 μm at this interface. A series of Ti-Al intermetallics may have been formed at this interfacial region according to Ti-Ag phase diagram [10].

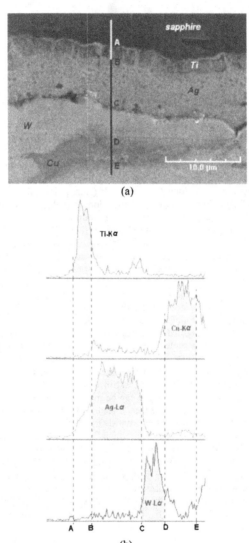

(a)

(b)

Figure 4. (a) SEM image of a sapphire/electrode interfacial area where the Ti and Ag coated sapphire piece is in contact with an elongated W particle in the W/Cu composite. (b) EDS line scan results at the locations illustrated by the vertical line in (a).

In contrast to the previous interfacial area where the metallized sapphire is in contact with an elongated W particle, the other type of interface where the metallized sapphire is adjacent to the Cu matrix part of W/Cu composite is also examined. The SEM image of such an interfacial area is shown in Figure 5 (a). Similarly, the dark portion at the top part of this figure represents sapphire. Adjacent to the sapphire is a Ti layer of about 3 μm thick with a column like structure (from point A and B in Figure 5 (a)). However, contrary to the existence of a distinctive interface between Ag coating and W particles as seen in the SEM image in Figure 4 (a), there is no clear-cut interface between Ag coating and Cu matrix is observed. Instead, a porous and homogeneous area is present (from point B and C in Figure 5 (a)) between the Ti coating and a colony of W particles. An EDS line scan was performed and the results are shown in Figure 5 (b). The most significant feature in this figure is the extensive overlapping of Cu-Kα and Ag-Lα peaks between point B and C, indicating a far-reaching elemental inter-diffusion in this region. The underlying process can best be described using the Ag-Cu phase diagram illustrated in Figure 6. As a eutectic type of system,

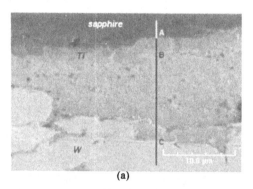

(a)

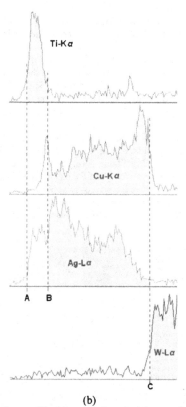

(b)

Figure 5. (a) SEM image of a sapphire/electrode interfacial area where the Ti and Ag coated sapphire piece is joined with Cu matrix in the W/Cu composite. (b) EDS line scan results at the locations illustrated by the vertical line in (a).

Ag and Cu exhibits limited solubility at temperatures below its eutectic point at 783°C. However, above this eutectic temperature the structural characteristics of a liquid-phase reaction were observed at the interface of a Cu-Ag bimetallic sample annealed at 800°C [11]. In another study, a eutectic composition of 39.9 at % Cu was detected using EDS at a Cu-Ag bimetallic interface [12]. In this study, it is believed that at the diffusion bonding temperature of 850°C a partial melting may have occurred at the interface between Ag coating and the Cu matrix, resulting in the disappearance of Ag-Cu interface, although three is no evidence of liquid phase reaction observed at the SEM magnifications examined. A strong bonding is expected in this type of interface due to the high degree of Ag-Cu inter-diffusion at the interface [13]. It is of interest to note that from Figure 5 (b) there is a significant overlapping of the Cu-Kα and the Ti-Kα peaks, as well as the Ag-Lα and Ti-Kα peaks between the locations A and B. This is an indication of a significant diffusion of Ag and Cu into the Ti coating.

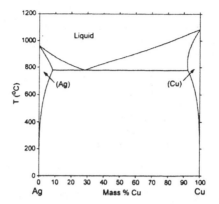

Figure 6. Ag-Cu phase diagram

The reaction between the Sapphire/Ti coating/Ag Coating/(W/Cu) was further investigated with nanoindentation testing. The hardness and the elastic modulus at the interface are roughly 2.5 GPa and 75 GPa respectively. In contrast the corresponding hardness and elastic modulus values within the copper phase are 4.5 GPa and 144 GPa and in sapphire phase are 36 GPa and 330 GPa. The soft interface thus provides a better cushion during thermal cycling.

CONCLUSIONS
The following conclusions can be drawn from this study:
- A novel method of joining sapphire to W/Cu metal surface using Directed Vapor Deposition (DVD) process has been demonstrated.
- Thin films of Ti and Ag have been effectively deposited onto the sapphire channel walls which have limited line-of-sight to the vapor source using DVD process.
- There is extensive overlapping of Ag-Lα and Cu-Kα peaks at the bonding interfacial region where the metallized sapphire is adjacent to the Cu matrix part of W/Cu composite, resulting in the disappearance of Ag-Cu interface. These results indicate a complete elemental inter-diffusion in this region.
- At the bonding interfacial areas where the metallized sapphire is in contact with the W particle part of the composite, no appreciable elemental inter-diffusion was observed.
- There is also a significant degree of overlapping of Ti-Kα and Ag-Lα peaks within the Ti coating, indicating comprehensive Ag diffusion into the Ti coating.
- The reaction zone between the Ti and sapphire could not be established with current characterization techniques. The nanoindentation work indirectly support interface reaction.

ACKNOWLEDGMENTS
The authors would like acknowledge the support of Prof. Xiaodong Li and Dr. Zhihui Xu for facilitating Nanoindentation effort at University of South Carolina, Columbia, SC. and Brian Muszynski, John Staudaher, Christina Elzey, and Brian Slawski of DVTI for assisting in various aspects of this experimental effort.

REFERENCES
[1] J.F. Groves, Y. Marciano, D.D. Hass, G. Mattausch, H. Morgner, H.N.G Wadley, Novel Vapor Phase Processing for High Performance Coatings, *Proceedings, Annual Technical Conference - Society of Vacuum Coaters*, 99-104 (2001).

[2] D.D. Hass, Y. Marciano, H.N.G. Wadley, Physical Vapor Deposition on Cylindrical Substrates, *Surface and Coatings Technology*, **185**, n. 2-3, July 22, 283-291 (2004).

[3] L.J. Kecskes, M.D. Trexler, B.R. Klotz, K.C. Cho, R.J. Dowding, Densification and Structural Change of Mechanically Alloyed W-Cu Composites, *Metallurgical and Materials Transactions A (Physical Metallurgy and Materials Science)*, **32A**, n. 11, 2885-93, November (2001).

[4] Akira Kanagawa, Jin-Quan Xu, Yoshiharu Mutoh, A Method for Improving Strength of Dissimilar Joints with Soft Metal by Cyclic Pre-loading, *Yosetsu Gakkai Ronbunshu/Quarterly Journal of the Japan Welding Society*, **21**, n 1, 68-72, February (2003).

[5] M.A. Butler and D.S. Ginley, Hydrogen Sensing with Palladium-Coated Optical Fibers, *J. of Appl. Phys.*, **64**, n.7, 3706-12, 1 October (1988).

[6] "Laser Brazing for Ceramic-to-Metal Joining", U.S. Patent # 5,407,119, April 18, 1995.

[7] Chiranjeev Kalra, Sohail H. Zaidi, Bruce J. Alderman, Richard B. Miles, and Y. V. Murty, Magnetically Driven Surface Discharges for Shock-Wave Induced Boundary-Layer Separation Control, *45th AIAA (American Institute of Aeronautics and Aestronautics), Aerospace Sciences Meeting and Exhibit, Reno, Nevada*, January 8-11, 2007.

[8] R. Jedamzik, A. Neubrand, J. Roedel, Characterisation of electrochemically processed graded tungsten/copper composites, *Materials Science Forum*, **308-311**, 782-787 (1999).

[9] J.F. Groves and H.N.G. Wadley, Functionally graded materials synthesis via low vacuum directed vapor deposition, *Composites Part B*, **28B**, 57-69 (1997).

[10] Max Hansen, "*Constitution of Binary Alloys*", McGraw Hill, 1958.

[11] A.G. Fitzgerald, H.L.L. Watton, and P.A. Moir, Microbeam Analysis Studies of the Copper-Silver Interface, *Journal of the Materials Science*, **28**, 1819-1823 (1993).

[12] P.A. Moir and A.G. Fitzgerald, in "EMAG-/MICRO'89", *Institute of Physics Conference Series*, **98** (IOP Publishing, Bristol, UK, 1990), p 315.

[13] L. Meng, S.P. Zhou, F.T. Yang, Q.J. Shen, M.S. Liu, Diffusion annealing of copper-Silver bimetallic strips at different temperatures, *Materials Characterization*, **47**, 269-274 (2001).

PROTECTIVE COATING ON METALS USING CHROMIUM-FREE ORGANIC-INORGANIC SILICA HYBRID AQUEOUS SOLUTION

Satomi Ono and Hiroyasu Tsuge
Nagoya Municipal Industrial Research Institute
3-4-41, Rokuban, Atsuta
Nagoya, Japan, 456-0058

ABSTRACT
    Thin organic-inorganic silica hybrid films providing good corrosion protection for metals were prepared using a chromium-free aqueous coating solution derived from silanes. The coating solution is prepared by optimizing the molar ratio of tetraethoxysilane, methyltriethoxysilane and acetic acid in water. The organic-inorganic silica hybrid films gave excellent corrosion protection for Zn-plating, stainless steel and aluminum alloy. Factors conferring improved corrosion protection are believed to be a chemical barrier caused by reaction of organic-inorganic silica hybrid with eluting metal cations, and a physical barrier due to dense packing of nano-particles made of the films.

1. INTRODUCTION

    Surface treatment on Zn-plating and on various metal products by hexavalent chromium conversion coating has been widely used, providing excellent corrosion protection for metals at low cost.   However the development of environmentally-friendly chromium-free aqueous reagents as chromate substitutes is a current urgent problem due to the European Directive (2000/53/EC) on ELV (End-of-Life Vehicles) and RoHS (European Restriction of the Use of Certain Hazardous Substances in Electrical and Electronic Equipment) regulations, which prohibit hexavalent chromium in scrapped cars, electric and electrical products.

    There are several alternatives to chromate treatments, including molybdenum phosphate chemical conversion, use of compounds of polymer and silane coupling agents, organic-inorganic hybrid films of polymer with phosphates, silica and metal compounds[1]. However, a practical chromium-free coating method providing adequate corrosion protection performance comparable to chromates has not yet been developed.

    We have asserted that the chemical solution deposition method is useful for developing chromium-free protective coatings[2, 3], since it is an energy conserving low cost technique capable of preparing homogeneous coatings on large areas of varying shape at relatively low temperatures.   To date, organic solvents have been used mainly in protective coatings by chemical solution deposition[2-14], but industry requires an organic solvent free non-flammable coating solution.   Accordingly, we describe the preparation of a coating solution from tetraethoxysilane, methyltriethoxysilane and acetic acid using water solvent, and investigate the factors that improve corrosion protection by the coatings of metals such as Zn-plating, stainless steel and aluminum alloy.

2. EXPERIMENTAL PROCEDURE

2.1 Preparation of coating films using silica aqueous solution

    Clear and homogeneous silica aqueous solutions were prepared by dissolving of tetraethoxysilane (TEOS) (Shin-Etsu Chemical Co., Ltd., Japan), methyltriethoxysilane (MTES) (Shin-Etsu Chemical Co., Ltd., Japan) and acetic acid (AcOH) into water in various ratios in the ambient atmosphere. The solutions were heated and concentrated at 70-80 °C to evaporate ethanol generated by hydrolysis of silanes. Coating films were prepared by dip-coating (withdrawal speed: 6.0 mm/s) on metals such as Zn-plated SPCC, stainless steel (SUS304) and aluminum alloy (A152S) after soaking in the solutions for 1 min. The films were dried at room temperature for 24 h.

2.2 Evaluation of coating films

The adhesion performance of the films prepared on metals was evaluated by the tape peeling test according to the Japanese Industrial Standard, JIS H 8504.

Morphology of the films was observed by scanning electron microscopy (SEM) (JSM-6300F, JEOL, Tokyo, Japan). It is difficult to observe the cross sections of the films on metals to determine the thickness by SEM because they are sub-micron thickness. The thickness of the films was estimated by observing the cross section of films prepared on silicon substrates under the same conditions as on metals. To determine the relation between the composition of the film and the factors giving corrosion protection, the surface of the films was analyzed by X-ray photoelectron spectroscopy (XPS) (AXIS-HSi, Shimadzu/KRATOS, Kyoto, Japan).

The corrosion protection performance of films on Zn-plating and A152S was evaluated by salt spray testing (SST) using 5 % NaCl solution at 35 °C according to JIS Z 2371. The presence of white zinc hydroxide rust was judged after 72 h spraying for the films prepared on Zn-plating, and the presence of white aluminum hydroxide rust was judged after 240 h spraying for the films prepared on A152S.

The corrosion rates of the films prepared on SUS304 were evaluated by the ferric chloride test according to JIS G 0578. The corrosion rate is defined as the weight loss per unit surface area per unit time after holding a coated specimen in 6% ferric chloride solution at 50 °C for 24h. In this paper, corrosion rates are calculated when the corrosion rate of the uncoated SUS304 reaches 100 %.

3. RESULTS AND DISCUSSION

3.1 Preparation of coating solution

High corrosion protection is never conferred by films prepared using aqueous solutions derived from the two components, TEOS and AcOH. However, films prepared using aqueous solutions and the three components TEOS, MTES and AcOH might improve the corrosion protection performance. The optimal ratio was investigated using solutions prepared with various ratios of TEOS, MTES and AcOH in water solvent.

In the silica aqueous solutions, the solution became turbid when the MTES content exceeded the TEOS in molar ratio. The compositions of all clear homogeneous solutions are summarized in Table I. The pHs of the solutions are in the interval 2.4-2.8. The films prepared on Zn-plating from solutions Nos.1-18 and No.23 are glossy or semi-glossy and show no exfoliation in the tape peeling test. The thickness of the films prepared using solutions No.5, No.7 or No.10 is 100-300 nm single coating on silicon substrates. The other films prepared on Zn-plating from solutions Nos.19-22 and Nos.24-27 are cloudy.

Table I. Corrosion protection performance of films prepared from silica aqueous solution.

No.	Total Conc. (mol/L)	Inorganic Comp. (%)	TEOS (mol%)	MTES (mol%)	AcOH (mol%)	pH	Film Appearance on Zn-plating	White Rust
1	1.0	73.3	55.0	20.0	25.0	2.7	Semi-glossy	No
2	1.0	78.6	55.0	15.0	30.0	2.6	Glossy	No
3	1.0	71.4	50.0	20.0	30.0	2.7	Glossy	No
4	1.0	76.9	50.0	15.0	35.0	2.6	Glossy	No
5	1.0	64.3	45.0	25.0	30.0	2.7	Glossy	No
6	1.0	69.2	45.0	20.0	35.0	2.7	Glossy	No
7	1.0	75.0	45.0	15.0	40.0	2.6	Glossy	No
8	1.1	77.8	43.75	12.5	43.75	2.6	Glossy	No
9	1.0	73.9	42.5	15.0	42.5	2.5	Glossy	No
10	1.0	66.7	40.0	20.0	40.0	2.6	Glossy	No
11	1.0	60.0	37.5	25.0	37.5	2.6	Glossy	No
12	1.2	75.0	37.5	12.5	50.0	2.5	Glossy	No
13	1.2	66.7	33.3	16.7	50.0	2.5	Glossy	No
14	1.0	80.0	60.0	15.0	25.0	2.8	Semi-glossy	Some
15	1.1	82.0	45.0	10.0	45.0	2.5	Semi-glossy	Some
16	1.2	80.0	40.0	10.0	50.0	2.5	Glossy	Some
17	1.3	66.7	30.0	15.0	55.0	2.4	Semi-glossy	Some
18	1.0	75.0	30.0	10.0	60.0	2.5	Semi-glossy	Some
19	1.0	75.0	60.0	20.0	20.0	2.8	Cloudy	Much
20	1.0	66.7	50.0	25.0	25.0	2.7	Cloudy	Much
21	1.0	84.6	50.0	10.0	40.0	2.6	Cloudy	Much
22	1.0	57.1	40.0	30.0	30.0	2.7	Cloudy	Much
23	1.1	87.5	46.7	6.6	46.7	2.5	Semi-glossy	Much
24	0.9	56.9	36.25	27.5	36.25	2.6	Cloudy	Much
25	0.9	50.0	33.3	33.3	33.4	2.6	Cloudy	Much
26	1.0	60.0	33.0	22.0	45.0	2.5	Cloudy	Much
27	1.4	75.0	33.3	11.1	55.6	2.4	Cloudy	Much

The corrosion protection of the films prepared on Zn-plating using solutions Nos.1-No27 in Table I is labelled by ○ (no white rust), △ (some white rust) and × (much white rust) in the triangular diagram in Figure 1. Figure 1 indicates the composition limits of the silica aqueous solutions for films giving high corrosion protection and no white rust on Zn-plating, as follows.
· 10 mol% < MTES ≤ 20 mol%,
  30 mol% < TEOS < 60 mol%, 25 mol% ≤ AcOH < 55 mol%
· 20 mol% < MTES ≤ 25 mol%,
  35 mol% < TEOS < 50 mol%, 30 mol% ≤ AcOH < 45 mol%

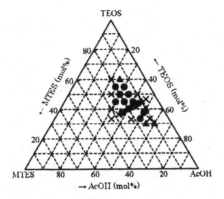

Figure 1. Composition of silica aqueous solution forming high corrosion resistant films on Zn-plating, ○: no white rust, Δ: some white rust and ×: much white rust.

Films with good adhesion and high corrosion protection performance on Zn-plating were prepared from the silica aqueous solutions in the appropriate composition zone. Stable clear solutions were never prepared when the MTES content exceeded that of TEOS. When the MTES content is too low, barrier activity of the films becomes inferior, resulting in lower corrosion protection performance. A higher AcOH content gives lower corrosion protection and too low content of AcOH leads to inferior stability of the solution and inferior film forming activity. The optimal AcOH content is approximately 30-40 mol%, and the pH is around 2.6. The solutions are stable for about 2 weeks at room temperature.

3.2 Composition and morphology of films prepared on Zn-plating

Table II shows the compositions of the films prepared from the solutions, a) 60:20:20 (No.19), b) 45:25:30 (No.5), c) 45:15:40 (No.7) and d) 30:10:60 (No.18) on Zn-plating before and after SST. Carbon, oxygen, silicon and zinc were detected by XPS analysis in all surfaces of these films. The presence of the zinc suggests that Zn_2SiO_4 is formed in the films as a result of the solubility of zinc in silica aqueous solution including AcOH. The chemical compounds in the films are consequently considered to be organic-inorganic silica hybrid (SiO_2 and $CH_3SiO_{3/2}$), zinc silicate (Zn_2SiO_4) and acetate residue (CH_3COO-) prior to SST. White rust consisting of zinc hydroxide ($Zn(OH)_2$) appears by corrosion after SST. Table III shows the contents of these chemical compounds, calculated using the atomic concentrations in Table II. No white rust appears in the films prepared from solutions, b) 45:25:30 and c) 45:15:40 in Table III. The content of the organic-inorganic silica hybrid is high in these films before SST. After SST, the content of the organic-inorganic silica hybrid decreases, and the content of the zinc silicate increases. No zinc hydroxide exists in the films.

White rust does appear in the films prepared from the solutions, a) 60:20:20 and d) 30:10:60 in Table III. There is little organic-inorganic silica hybrid and a high zinc silicate content in the films before SST. After SST, organic-inorganic silica hybrid has disappeared and the zinc silicate and the zinc hydroxide content increases.

Table II. Composition of films prepared on Zn-plating before and after SST.

	TEOS (mol%)	MTES (mol%)	AcOH (mol%)	C (mol%)		O (mol%)		Si (mol%)		Zn (mol%)	
				Before SST	After SST	Before SST	After SST	Before SST	After SST	Before SST	After SST
a)	60	20	20	18	22	55	51	15	5	12	22
b)	45	25	30	20	18	53	52	18	13	9	17
c)	45	15	40	24	17	53	54	16	11	8	18
d)	30	10	60	24	18	51	53	12	7	13	23

Errors: within ±10%

Table III. Contents of chemical compounds of films prepared on Zn-plating before and after SST.

	TEOS (mol%)	MTES (mol%)	AcOH (mol%)	SiO_2 $(CH_3SiO_{3/2})$ (mol%)		Zn_2SiO_4 (mol%)		$Zn(OH)_2$ (mol%)		Acetate Residues (mol%)	
				Before SST	After SST	Before SST	After SST	Before SST	After SST	Before SST	After SST
a)	60	20	20	42	0	28	25	0	75	30	0
b)	45	25	30	61	26	21	48	0	0	18	26
c)	45	15	40	53	11	18	50	0	0	29	39
d)	30	10	60	48	0	33	36	0	46	19	18

The results indicate that the organic-inorganic silica hybrid and zinc silicate composite, which is formed by reaction of the organic-inorganic silica hybrid with eluting zinc cations, prevents the generation of white rust and improves corrosion protection.

Figure 2 shows SEM images of the films prepared on Zn-plating before and after SST. Before SST, the film prepared from solution a) 60:20:20 shows large cracks, whereas the films prepared from solutions b) 45:25:30, c) 45:15:40 and d) 30:10:60 are smooth and dense with small cracks. The film prepared from solution a) 60:20:20, in which much white rust is observed, shows a sponge-like texture due to corrosion after SST. In contrast, the film prepared from solution b) 45:25:30 shows no white rust, and no difference is observed before and after SST in SEM images of the films. In the film prepared from solution c) 45:15:40, some particles of sub-micron size were observed in the surface after SST, in which no white rust is observed. Some petal-like corrosion products were observed in the film prepared from solution a) 30:10:60 after SST, in which some white rust appears.

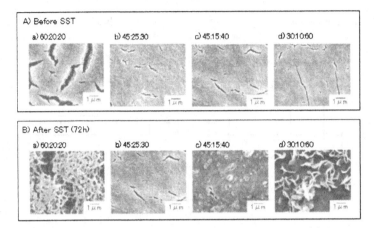

Figure 2. SEM images of films prepared on Zn-plating before and after SST, a) 60:20:20, b) 45:25:30, c) 45:15:40 and d) 30:10:60.

Table IV. Effect of the concentration of silica aqueous solution (45:25:30) on appearance and corrosion protection performance of films prepared on Zn-plating.

Coating Solution TEOS : MTES : AcOH = 45 : 25 : 30 (mol/L)	Total Conc. (mol/L)	pH	Film Surface Appearance	
			Before SST	After SST (72h) White Rust
a) TEOS 0.45 MTES 0.25 AcOH 0.30	1.0	2.7	Glossy	No
b) TEOS 0.39 MTES 0.21 AcOH 0.26	0.86	2.7	Glossy	No
c) TEOS 0.32 MTES 0.18 AcOH 0.21	0.71	2.7	Semi-glossy	Some
d) TEOS 0.26 MTES 0.14 AcOH 0.17	0.57	2.8	Cloudy	Much

Table IV shows the effect of the concentration of the solution (45:25:30) on the appearance and the corrosion protection performance of the films prepared on Zn-plating. The films were prepared from various concentrations of the solutions, a) 1.0 mol/L, b) 0.86 mol/L, c) 0.71 mol/L and d) 0.57 mol/L. The films prepared from solutions a) 1.0 mol/L and b) 0.86

mol/L appeared glossy before SST, and no white rust was observed after SST. The film prepared from solution c) 0.71 mol/L appeared semi-glossy before SST, and some white rust was observed after SST. The film prepared from solution d) 0.57 mol/L was cloudy before SST, and much white rust was observed after SST.

Figure 3 shows SEM images of the films before and after SST. There is no difference in the surfaces of the films prepared from solutions a) 1.0 mol/L and b) 0.86 mol/L before and after SST. on which no white rust is observed. The films prepared from solutions c) 0.71 mol/L and d) 0.57 mol/L show corrosion texture after SST, and much white rust is observed.

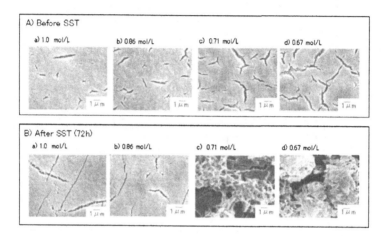

Figure 3. SEM images of films prepared from silica aqueous solution (45:25:30) on Zn-plating before and after SST, a) 1.0 mol/L, b) 0.86 mol/L, c) 0.71 mol/L and d) 0.57 mol/L.

Table V shows the compositions of the films, and Table VI summarizes the content of the chemical compounds in the films before and after SST.

The films prepared from solutions, a) 1.0 mol/L and b) 0.86 mol/L reveal that the content of the organic-inorganic silica hybrid is high before SST and decreases, whereas the zinc silicate content increases after SST. The film prepared from solution c) 0.71 mol/L shows that the organic-inorganic silica hybrid which exists before SST has disappeared and zinc silicate and zinc hydroxide exist in higher content after SST. The film prepared from solution d) 0.57 mol/L shows that the organic-inorganic silica hybrid and zinc silicate, which are present before SST have disappeared and the zinc hydroxide content reaches 100%.

Table V. Composition of films prepared from silica aqueous solution (45:25:30) on Zn-plating before and after SST.

No.	TEOS (mol/L)	MTES (mol/L)	AcOH (mol/L)	C (mol%)		O (mol%)		Si (mol%)		Zn (mol%)	
				Before SST	After SST	Before SST	After SST	Before SST	After SST	Before SST	After SST
1	0.45	0.25	0.30	20	18	53	52	18	13	9	17
2	0.39	0.21	0.26	19	18	56	52	17	10	8	20
3	0.32	0.18	0.21	20	21	54	51	16	6	10	22
4	0.26	0.14	0.17	19	20	54	51	16	0	11	29

Errors: within ± 7%

Table VI. Content of chemical compounds of films prepared from silica aqueous solution (45:25:30) on Zn-plating before and after SST.

No.	TEOS (mol/L)	MTES (mol/L)	AcOH (mol/L)	$SiO_2 (CH_3SiO_{3/2})$ (mol%)		Zn_2SiO_4 (mol%)		$Zn(OH)_2$ (mol%)		Acetate Residues (mol%)	
				Before SST	After SST	Before SST	After SST	Before SST	After SST	Before SST	After SST
1	0.45	0.25	0.30	61	26	21	48	0	0	18	26
2	0.39	0.21	0.26	54	0	17	63	0	0	29	37
3	0.32	0.18	0.21	50	0	23	31	0	51	27	18
4	0.26	0.14	0.17	50	0	25	0	0	100	25	0

These results indicate that the silica aqueous solution must have concentration more than 0.86 mol/L to generate films with good corrosion protection. This is a result of the reactivity of the acidic coating solution with zinc. Low acidity in the low concentration solution is insufficient to form dense films of organic-inorganic silica hybrid and zinc silicate composite nano-particles by chemical-conversion. The factors that improve corrosion protection on Zn-plating are believe to be a chemical barrier caused by reaction of organic-inorganic silica hybrid with eluting zinc cations and a physical barrier due to dense packing of the organic-inorganic silica hybrid and zinc silicate composite nano-particles.

SST, XPS analysis and SEM observation indicate the film prepared from the solution with composition 45:25:30 has the best corrosion protection performance. We suggest that good corrosion protection is gained by the formation of dense film composed of nano-composites with more than 50 mol% of the organic-inorganic silica hybrid and less than 21 mol% zinc silicate.

3.3 Evaluation of films prepared on SUS304 and Al52S

Corrosion protection was investigated for films prepared on SUS304 and Al52S using various concentrations of the solutions, a) 1.0 mol/L, b) 0.86 mol/L, c) 0.71 mol/L and d) 0.57

mol/L in the composition 45:25:30. Table VII shows the corrosion rates of films on SUS304 evaluated by the ferric chloride test. The corrosion rates were in the range of 17-24 %. The films prepared by solution c) 0.71 mol/L confer the highest corrosion protection among these concentrations. Figure 4 shows SEM images of the films prepared on SUS304, a) 1.0 mol/L and b) 0.71mol/L. The films are composed of close packed nano-particles and are free from defects such as cracks and pores, unlike films on Zn-plating.

XPS analysis revealed carbon, oxygen and silicon in the films on SUS304. The result indicates that the films prepared on SUS304 are composed of the organic-inorganic silica hybrid and acetates, because SUS304 has no reactivity with the acidic coating solution at pH 2-3. Good corrosion performance of the film does not therefore depend on the concentration of the coating solution. The films after the ferric chloride test were not analyzed by XPS because they had peeled off after the test.

Table VII. Corrosion rates of films prepared from silica aqueous solution (45:25:30) on SUS304.

Coating Solution TEOS : MTES : AcOH = 45 : 25 : 30 (mol/L)	Corrosion Rates* (%)
a) TEOS: 0.45, MTES: 0.25, AcOH: 0.30	20 ± 5
b) TEOS: 0.39, MTES: 0.21, AcOH: 0.26	19 ± 2
c) TEOS: 0.32, MTES: 0.18, AcOH: 0.21	17 ± 1
d) TEOS: 0.26, MTES: 0.14, AcOH: 0.17	24 ± 1

*Corrosion rate of non-coated SUS304=100%

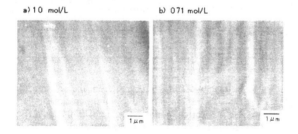

a) 1.0 mol/L   b) 0.71 mol/L

Figure 4. SEM images of films prepared from silica aqueous solution (45:25:30) on SUS304, a) 1.0 mol/L and b) 0.71mol/L.

The films prepared on Al52S also gave high corrosion protection with no white rust after SST in these solution concentrations. Table VIII shows the composition of film prepared from c)

0.71 mol/L solution (45:25:30) on Al52S before and after SST. Carbon, oxygen and silicon were detected in the films before SST and, in addition to these atoms, aluminum was detected in the films after SST.

Although Al52S is considered to react with acidic solution, aluminum was not detected in the surface of the film. The results indicate that the films are formed from the solution with reaction only in the interface of the film and Al52S before SST, and that corrosion protection is enhanced by generating insoluble aluminum silicate (Al_2SiO_5) as a result of the reaction of organic-inorganic silica hybrid with eluted aluminum cations after SST. The content of chemical compounds was calculated using the atomic concentrations in Table VIII. The results indicate that 14mol% aluminum silicate is formed after SST (Table VIIII).

Table VIII. Composition of film prepared from silica aqueous solution (45:25:30) on Al52S before and after SST.

TEOS (mol/L)	MTES (mol/L)	AcOH (mol/L)	C (mol%)		O (mol%)		Si (mol%)		Al (mol%)	
			Before SST	After SST	Before SST	After SST	Before SST	After SST	Before SST	After SST
0.32	0.18	0.21	20	26	58	54	22	13	0	7

Table VIIII. Content of chemical compounds of film prepared from silica aqueous solution (45:25:30) on Al52S before and after SST.

TEOS (mol/L)	MTES (mol/L)	AcOH (mo/L)	SiO_2 ($CH_3SiO_{3/2}$) (mol%)		Al_2SiO_5 (mol%)		Acetate Residues (mol%)	
			Before SST	After SST	Before SST	After SST	Before SST	After SST
0.32	0.18	0.21	76	45	0	14	24	41

Figure 5 shows SEM images of films prepared from the solution (45:25:30) on Al52S before and after SST. The films appear smooth and free from cracks, similar to that on SUS304 (Fig. 4). These results suggest that corrosion protection on SUS304 and Al52S is improved by a chemical barrier caused by reaction of organic-inorganic silica hybrid with eluting metal cations, and a physical barrier due to close packing of nano-particles in organic-inorganic silica hybrid.

4. CONCLUSIONS

The corrosion protection performance of metals was improved by coating of organic-inorganic silica hybrid films prepared from a solution having an optimum ratio of TEOS, MTES and AcOH in water. In the coating on Zn-plating, organic-inorganic silica hybrid and zinc silicate constitute nano-composites as a result of the solubility of zinc in acidic coating solution that includes AcOH. Improved corrosion protection on Zn-plating is believed to be due to a chemical barrier caused by reaction of organic-inorganic silica hybrid with eluting zinc cations, and a physical barrier set up by compacting of nano-particles in the organic-inorganic silica hybrid and zinc silicate composite. SUS304 is not reactive and are insoluble in the acid coating solution and Al52S might reactive with acidic solution only in the interface of the film and substrate. The films are therefore composed of a dense organic-inorganic silica hybrid. Corrosion protection on SUS304 and Al52S is believed to be due to a chemical barrier caused by reaction of organic-inorganic silica hybrid with eluting metal cations, and a physical barrier caused by the compacting of nano-particles in organic-inorganic silica hybrid.

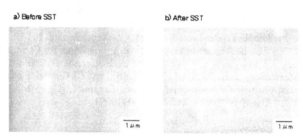

Figure 5. SEM images of films prepared from silica aqueous solution (45:25:30) on Al52S, a) before SST and b) after SST.

REFERENCES

[1]T. Aoe, "Replacements for Chromate Conversion Coatings" *Hyomen Gijyutu*, in Japanese, **49**, 221-29 (1998).

[2]S. Ono, Y. Nishi and S. Hirano, "Chromium-Free Corrosion Resistance of Metals by Ceramic Coating" *J. Am. Ceram. Soc.*, **84** 3054 (2001).

[3]S. Ono, H. Tsuge, Y. Nishi and S. Hirano, "Improvement of corrosion resistance of metals by an environmentally friendly silica coating method" *J. Sol-Gel Sci. Technol.*, **29**, 147-153 (2004).

[4]O. de Sanctis, L. Gómez, N. Pellegri, C. Parodi, A. Marajofsky and A. Durán, "Protective Glass Coatings on Metallic Substrates" *J. Non-Cryst. Solids*, **121**, 338-41 (1990).

[5]K. Izumi, H. Tanaka, M. Murakami, T. Deguchi, A. Morita, N. Tohge and T. Minami, "Coating of Fluorine-doped ZrO_2 Films on Steels by Sol-Gel Method" *J. Non-Cryst. Solids*, **121**, 344-47 (1990).

[6]A. R. Di Giampaolo Conde, M. Puerta, H. Ruiz and J. Lira Olivares, "Thick Aluminosilicate Coatings on Carbon Steel via Sol-Gel" *J. Non-Cryst. Solids*, **147&148**, 467-73 (1992).

[7]K. Izumi, H. Tanaka, Y. Uchida, N. Tohge and T. Minami, "Influence of Firing Conditions on Adhesion of Methyltrialkoxysilane-Derived Coatings on Steel Sheets" *J. Non-Cryst. Solids*, **147&148**, 483-87 (1992).

[8]M. Atik, J. Zarzycki and C. R'kha, "Protection of Ferritic Stainless Steel Against Oxidation by Zirconia Coatings" *J. Mater. Sci. Lett.*, **13**, 266-69 (1994).

[9]M. Atik, C. R'kha, P. De Lima Neto, L. A. Avaca, M. A. Aegerter and J. Zarzycki, "Protection of 316L Stainless Steel by Zirconia Sol-Gel Coatings in 15% H_2SO_4 Solutions" *J. Mater. Sci. Lett.*, **14**, 178-81 (1995).

[10]M. F. M. Zwinkels, S. G. Järås, P. G. Menon and K. I. Åsen, "Preparation of Anchored Ceramic Coatings on Metal Substrates: A Modified Sol-Gel Technique Using Colloidal Silica Sol" *J. Mater. Sci.*, **31**, 6345-49 (1996).

[11]S. Hirai, K. Shimakage, M. Sekiguchi, K. Wada and A. Nukui, "Zirconium Oxide Coating on Anodized Aluminum by the Sol-Gel Process Combined with Ultraviolet Irradiation at Ambient Temperature" *J. Am. Ceram. Soc.*, **82**, 2011-16 (1999).

[12]P. B. Kirk and R. M. Pilliar, "The Deformation Response of Sol-Gel-Derived Zirconia Thin Films on 316L Stainless Steel Substrates Using a Substrate Straining Test" *J. Mater. Sci.*, **34**, 3967-75 (1999).

[13]L. F. . Perdomol, P. De Lima-Neto, M. A. Aegerter and L. A. Avaca, "Sol-Gel Deposition of ZrO_2 Films in Air and in Oxygen-Free Atmospheres for Chemical Protection of 304 Stainless Steel: A Comparative Corrosion Study" *J. Sol-Gel Sci. Technol.*, **15**, 87-91 (1999).

[14]S. H. Messaddeq, S. H. Pulcinelli, C.V. Santilli, A. C. Guastaldi and Y. Messaddeq, "Microstructure and Corrosion Resistance of Inorganic-Organic (ZrO_2-PMMA) Hybrid Coating on Stainless Steel" *J. Non-Cryst. Solids*, **247**, 164-170 (1999).

AN ENERGY MODEL OF SEGMENTATION CRACKING OF SiO$_x$ THIN FILM ON A POLYMER SUBSTRATE

Marcin Białas, Zenon Mróz
Institute of Fundamental Technological Research, Polish Academy of Sciences
Świętokrzyska 21 St
00-049 Warsaw, Poland

ABSTRACT

In order to measure the fracture energy of a silicon oxide thin film deposited on a poly(ethylene terephthalate)(PET) substrate an energy model of film segmentation cracking has been formulated. The model focuses on a topological transformation between an intact and a damaged structure, rather than on a constitutive modeling of fracture itself. An energy transition condition should be satisfied in order for cracks to occur.

INTRODUCTION

Progress in thin film and polymer technology has made it possible to deposit ceramic films of thickness in the range of 10÷100 nm on flexible polymer substrates. The total thickness of these composites varies from 10 to 100 μm. Industrial applications are SiO$_x$ (x~1.65) films deposited by chemical or physical vapor deposition processes on polyethylene terephthalate (PET) substrates. They found a considerable interest for packaging purposes, specifically in the pharmaceutical and food industries[1-9]. In such applications, the coating integrity should be assured during the whole life-time of a package. In order to characterize the fracture strength of the film, a quantitative method of estimating the resistance to tensile loadings should be established. The technique of multiple film cracking has been used[1-9], where the relation between the measured mean crack spacing and the applied strain was used to characterize the film strength or fracture energy.

The present paper provides an analysis of delamination of a thin brittle film deposited on a stretched substrate. A constant shear stress at the fully damaged film/substrate interface is assumed. This assumption takes into account slip mechanisms taking place at the film/substrate interface. The interfacial shearing responsible for debonding is a result of slip deformation mechanisms observed for PET at high strains in small-angle x-ray spectrometry measurements[10]. In the papers by Leterrier et al.[4-5] the substrate shear stress at saturation is compared with the interfacial shear strength, proving that up to 120°C both are equivalent within the experimental scatter. Moreover, the adhesive strength was found to be independent of coating's thickness. Yanaka et al.[7] observed that the crack spacing within SiO$_x$ film on PET substrate was proportional to the film thickness. They argued that such a behaviour is expected for a constant shear stress at the film/substrate interface. On the basis of this analysis an energy model of film segmentation cracking is formulated. A transition condition states that the potential and dissipated energy of the cracked film cannot be greater than the energy of the intact structure and gives rise to a topological transformation of the coating. In order to validate the model, the fracture energy of SiO$_x$ film deposited on PET substrate was calculated, assuming the crack spacing and other mechanical and geometrical parameters of the system as given.

PROBLEM FORMULATION

Let us consider a composite plate consisting of a substrate of width w and length L with a deposited thin film as presented in Figure 1(a). External loading σ_0 is subjected to the substrate material causing its elongation. In the course of loading the film can delaminate and segmentation cracking takes place within the coating perpendicular to the direction of the applied loading. In order to treat the problem analytically we shall consider a simple one dimensional strip model, where the interaction between the substrate and the film is realized by an interface of zero thickness, see Figure 1(b). The main stress components are normal stresses σ_f and σ_s, respectively within the film and the substrate. The interaction between the composite constituents is executed by the shear stress τ at the film/substrate interface. We assume the relation between the stress τ and the relative displacement at the interface presented in Figure 2 – a constant shear stress τ remains independent of the actual value of the relative displacement. The equilibrium equations for the film and the substrate have the forms

$$\frac{d\sigma_f}{dx} - \frac{\tau}{h_f} = 0 \tag{1}$$

$$\frac{d\sigma_s}{dx} + \frac{\tau}{h_s} = 0 \tag{2}$$

Equations (1) and (2) can be obtained by considering an equilibrium state of an infinitesimal fragment of the composite as presented in Figure 2. The heights of the film and the substrate are denoted respectively by h_f and h_s.

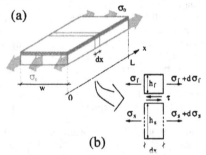

Figure 1. (a) Thin film deposited on a stretched substrate. (b) One dimensional strip model.

Both the film and the substrate are elastic, so we can write

$$\varepsilon_f = \frac{\sigma_f}{E_f} \tag{3}$$

$$\varepsilon_s = \frac{\sigma_s}{E_s} \tag{4}$$

An Energy Model of Segmentation Cracking of SiO$_x$ Thin Film on a Polymer Substrate

Young moduli are denoted by E_f and E_s, elongations by ε_f, ε_s with respect to the film and to the substrate. To fully formulate the governing equations we write the formulas relating strain ε_f and ε_s with the displacement fields u_f and u_s along the x axis, respectively for the film and the substrate. They take the form

$$\frac{du_f}{dx} = \varepsilon_f \tag{5}$$

$$\frac{du_s}{dx} = \varepsilon_s \tag{6}$$

Equations (1)÷(6) together with the interface constitutive law

$$\dot{u}_f - \dot{u}_s = \dot{\lambda} sign(\tau), \quad \dot{\lambda} \geq 0, \quad |\tau| - \tau_c \leq 0, \quad \dot{\lambda}(|\tau| - \tau_c) = 0 \tag{7}$$

provide slip conditions and will serve as a basis for an energy model of segmentation cracking within the film.

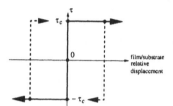

Figure 2. Interface constitutive law.

FILM DELAMINATION

The external loading is subjected to the substrate and causes the elongation of the whole system. The tensional stresses are transmitted to the film from the substrate through the interfacial shear stresses, resulting from film/substrate relative slip. The configuration of shear and normal stresses at this stage is presented in Figure 3. We can see delamination zones developing from both ends of the composite with its middle section remaining fully bonded. The length of a single delamination zone is indicated by d. The resulting film and substrate stress and strains can be obtained from equations (1), (2), (3), (4) and (7). In the film they are described by the functions

$$\sigma_f(x) = \begin{cases} \dfrac{\tau_c}{h_f} x & x \in <0, d> \\[2mm] \dfrac{\tau_c}{h_f} d & x \in <d, L-d> \end{cases} \tag{8}$$

$$\varepsilon_f(x) = \begin{cases} \dfrac{\tau_c}{h_f E_f} x & x \in <0, d> \\[2ex] \dfrac{\tau_c}{h_f E_f} d & x \in <d, L-d> \end{cases} \tag{9}$$

The boundary condition $\sigma_f(0)=0$ have been imposed to obtain (8) and (9). The fields $\sigma_f(x)$ and $\varepsilon_f(x)$ for $x \in <L-d, L>$ have to be symmetric with respect to the symmetry axis of the structure. The distribution of $\sigma_f(x)$ is schematically presented in Figure 3(b).

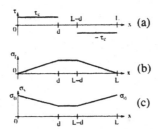

Figure 3. Film delamination: (a) shear stresses at the interface; (b) film normal stress; (c) substrate normal stress.

By imposing the boundary condition $\sigma_s(0)=\sigma_0$ we obtain the stress σ_s and the strain ε_s in the substrate

$$\sigma_s(x) = \begin{cases} \sigma_0 - \dfrac{\tau_c}{h_s} x & x \in <0, d> \\[2ex] \sigma_0 - \dfrac{\tau_c}{h_s} d & x \in <d, L-d> \end{cases} \tag{10}$$

$$\varepsilon_s(x) = \begin{cases} \dfrac{\sigma_0}{E_s} - \dfrac{\tau_c}{h_s E_s} x & x \in <0, d> \\[2ex] \dfrac{\sigma_0}{E_s} - \dfrac{\tau_c}{h_s E_s} d & x \in <d, L-d> \end{cases} \tag{11}$$

The fields $\sigma_s(x)$ and $\varepsilon_s(x)$ for $x \in <L-d, L>$ have to be symmetric with respect to the symmetry axis of the structure. The distribution of $\sigma_s(x)$ is schematically presented in Figure 3(c).

In order to obtain the relation between the traction σ_0 and the length of a single damage zone d we have to impose the condition $\varepsilon_s=\varepsilon_f$ for $x \in <d, L-d>$, meaning that there is no relative displacement between the film and the substrate in this particular area. By using equations (9) and (11) we obtain

$$\sigma_0 = \frac{\tau_c (E_s h_s + E_f h_f)}{E_f h_f h_s} d \tag{12}$$

For $d=L/2$ the whole interface is damaged. Due to the constant value of the interface stress $\tau=\tau_c$ any further increase in the loading traction σ_0 subjected to the substrate does not influence the stress distribution within the film. The film stress can be obtained from equation (8) for $d=L/2$.

Let us derive now the formulas for the displacement fields of the substrate and of the film for $x \in <0,d>$, that is for the zone where the substrate and the film contact through a damaged interface. They will be used in the subsequent section, where an energy model of film segmentation cracking is formulated.

By integrating function (11) for $x \in <0,d>$ and using formula (12) one obtains the displacement field of the substrate within the region $x \in <0,d>$

$$u_s(x) = u_s(0) + \frac{x \left[4d \left(E_f h_f + E_s h_s \right) - E_f h_f x \right] \tau_c}{2 E_f E_s h_f h_s} \tag{13}$$

where $u_s(0)$ is the integration constant being the value of $u_s(x)$ for $x=0$. Assuming that the substrate is constrained for $x=0$, $u_s(0)=0$ we obtain from equation (13) the displacement of the substrate material for $x=L$ in the form:

$$u_s(L) = \frac{\left(d^2 E_f h_f + E_s h_s L d \right) \tau_c}{E_f E_s h_f h_s} \tag{14}$$

Integration of function (9) for $x \in <0,d>$ with the continuity condition $u_f(d)=u_s(d)$ provides the displacement field of the film within the region $x \in <0,d>$

$$u_f(x) = u_s(0) + \frac{\left[d^2 \left(E_f h_f + E_s h_s \right) + E_s h_s x^2 \right] \tau_c}{2 E_f E_s h_f h_s} \tag{15}$$

Here again equation (12) has been used.

The system response to the subjected tensile loading has been fully described. The assumed one dimensional strip model allowed for analytical expressions of all physical quantities. Now we can proceed to the main subject of this contribution being an energy model of film segmentation cracking.

ENERGY MODEL

The energy of an uncracked composite is expressed by the formula

$$\Pi = \Sigma_s + \Sigma_f + D_t - \Pi_\sigma \tag{16}$$

where

$$\Sigma_f = 2 \left(\frac{1}{2} w h_f \int_0^d \frac{\sigma_f(x)^2}{E_f} dx \right) + \frac{1}{2} w h_f \frac{\sigma_f^2}{E_f} (L - 2d) \tag{17}$$

is the elastic energy stored in the film;

$$\Sigma_s = 2\left(\frac{1}{2}wh_s\int_0^d\frac{\sigma_s(x)^2}{E_s}dx\right)+\frac{1}{2}wh_s\frac{\sigma_s^2}{E_s}(L-2d) \qquad (18)$$

is the elastic energy stored in the substrate:

$$D_r = -2w\tau_c\int_0^d\left(u_s(x)-u_f(x)\right)dx \qquad (19)$$

is the energy dissipated on the film/substrate interface by the shear stress;

$$\Pi_\sigma = \sigma_0 h_s w u_s(L) \qquad (20)$$

is the potential of external forces acting on the substrate. Thus, the energy Π should be understood as the potential energy of the structure plus the energy dissipated at the film/substrate interface.

The first terms in equations (17) and (18) refer to the elastic energy stored within the film and the substrate contacting through two delaminated zones and $\sigma_f(x)$ and $\sigma_s(x)$ are provided by equations (8) and (10), respectively. The second terms in (17) and (18) refer to the elastic energy stored within the film and the substrate perfectly bonded with each other and, according to equations (8) and (10), we have $\sigma_f=\tau_c d/h_f$, $\sigma_s=\sigma_0-\tau_c d/h_s$. By substituting (8), (10), (12) (13), (14) and (15) into (16) we obtain

$$\Pi = -\frac{\left(2d^3E_fh_f+3E_sh_sLd^2\right)}{6E_f^2E_sh_f^2h_s}(E_fh_f+E_sh_s)w\tau_c^2 \qquad (21)$$

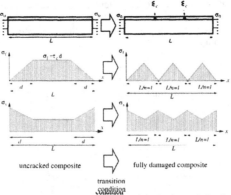

Figure 4. Topological transformation from an uncracked to a fully damaged structure.

Let us consider now a composite consisting of n film fragments contacting with the substrate through a fully damaged interface as presented in Figure 4. The system energy for this particular case can be expressed by the formula

$$\Pi'' = \Sigma_s'' + \Sigma_f'' + D_\tau'' - \Pi_\sigma'' + (n-1)wh_f\Sigma_c \tag{22}$$

where

$$\Sigma_f'' = 2\left(\frac{1}{2}nwh_f \int_0^{L/(2n)} \frac{\sigma_f(x)^2}{E_f}\,dx\right) \tag{23}$$

is the elastic energy stored in n film fragments;

$$\Sigma_s'' = 2\left(\frac{1}{2}nwh_s \int_0^{L/(2n)} \frac{\sigma_s(x)^2}{E_s}\,dx\right) \tag{24}$$

is the elastic energy stored in the substrate;

$$D_\tau'' = -2nw\tau_c \int_0^{L/(2n)} \left(u_s(x) - u_f(x)\right)dx \tag{25}$$

is the energy dissipated at the fully damaged interface by n film fragments contacting with the substrate;

$$\Pi_\sigma'' = \sigma_0 h_s w u_s''(L) \tag{26}$$

is the potential of external forces acting on the substrate. The last term in equation (22) refers to n-1 cracks formed, with Σ_c being the specific energy of a unit area. The term $(n-1)wh_f\Sigma_c$ is the energy dissipated to form n-1 segmentation cracks. Similar to the previous case, the energy Π'' is the potential energy of the composite plus the dissipated energy. In the present case, however, the dissipated energy is the energy dissipated at the film/substrate interface and the energy dissipated to create segmentation cracks.

According to equations (8) and (10) we have

$$\sigma_f(x) = \frac{\tau_c}{h_f}x, \qquad \sigma_s(x) = \sigma_0 - \frac{\tau_c}{h_s}x \tag{27}$$

in equations (23) and (24). The displacements $u_s(x)$ and $u_f(x)$ in (25) can be specified for the present case by setting $d=L/(2n)$ in equations (13) and (15). We obtain

An Energy Model of Segmentation Cracking of SiO$_x$ Thin Film on a Polymer Substrate

$$u_s(x) = u_s(0) + \frac{x\left[2L(E_f h_f + E_s h_s)/n - E_f h_f x\right]\tau_c}{2E_f E_s h_f h_s}$$ (28)

$$u_f(x) = u_s(0) + \frac{\left[L^2(E_f h_f + E_s h_s)/(4n^2) + E_s h_s x^2\right]\tau_c}{2E_f E_s h_f h_s}$$ (29)

The substrate displacement $u_s''(L)$ in equation (26) can be expressed by

$$u_s''(L) = 2n \int_0^{L/(2n)} \varepsilon_s(x)dx$$ (30)

with $\varepsilon_s(x) = \sigma_0/E_s - \tau_c x/(h_s E_s)$ according to equation 11(a).

For the fragmentation process to occur we postulate the following transition condition

$$\Pi \geq \Pi''$$ (31)

stating, that the energy Π'' of a damaged system cannot be greater than the energy Π of the intact structure. This should be satisfied for the value of external loading σ_0 identical for the uncracked and the fully damaged composite, as presented in Figure 4. Inequality (31) gives a physical meaning to a transformation changing the topology of the intact film into a system of n segments forming a maximally damaged structure. This approach does not account for a loading history and considers only energy levels of the intact and the fully fragmented system. The condition (31) states that the cracks will occur when the energy Π of an intact composite happens to be greater than the energy Π'' of a cracked structure consisting of n fragments. Bearing in mind that the energy Π is an increasing function of d, that is the length of a friction zone, we can formulate what follows. The cracks will occur when for increasing d the inequality (31) happens to be true. The consequent number of strip fragments n is a function of the smallest possible length d satisfying the transition condition (31) and, as an unknown parameter, can be derived from (31).

By substituting (12), (27), (28), (29) and (30) into (22) we obtain

$$\Pi'' = \frac{E_f h_f L^2 - 6d(E_f h_f d + E_s h_s d)n^2}{12 E_f^2 E_s h_f^2 h_s n^2}(E_f h_f + E_s h_s)Lw\tau_c^2 + (n-1)wh_f\Sigma_c$$ (32)

Using equations (21) and (32) the inequality (31) can be rewritten now in the following form

$$\frac{(\lambda+1)\left[\xi^2(3-\xi)n^2 - 2\right]}{(n-1)n^2} \geq \beta$$ (33)

where the dimensionless parameters have been introduced

$$\xi = \frac{d}{L/2}, \qquad \lambda = \frac{E_f h_f}{E_s h_s}, \qquad \beta = \frac{24 h_f^2 E_f}{L^3 \tau_c^2}\Sigma_c$$ (34)

Thus, ζ is a dimensionless length of a single damaged zone, $0 \leq \zeta \leq 1$. By introducing

$$\beta_n(\xi) = \frac{(\lambda+1)\left[\xi^2(3-\xi)n^2 - 2\right]}{(n-1)n^2} \tag{35}$$

we can write

$$\beta_n(\xi) \geq \beta \tag{36}$$

According to what has been stated above, condition (36) has to be satisfied for the smallest possible length of friction zone, that is for the smallest possible ζ. Condition (36) states that the elastic energy of the intact structure is big enough when compared to the specific surface energy of cracks. Thus, the cracks will occur when the increasing elastic energy is high enough to be used as a driving force for the fragmentation process. Formula (35) provides a whole family of functions $\beta_n(\xi)$. They are presented in Figure 5 for $\lambda=0$ and for several values of n.

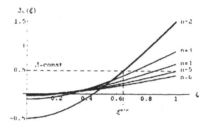

Figure 5. The plots of functions $\beta_n(\xi)$ for $\lambda=0$.

Let us have an equality in the condition (36) and assume that β is a given value defined by the mechanical and geometrical parameters of the system. The smallest possible ζ^{min} that satisfies $\beta=\beta_n(\zeta)$ for the whole family of functions β_n is the actual length of a single delamination zone leading to the segmentation process. The parameter n for which $\beta=\beta_n(\zeta^{min})=$const happens to be true, is the actual number of strip fragments occurring for a system characterized by the dimensionless parameter β. The parameters ζ^{min} and n can be obtained from Figure 5 by considering an intersection of line $\beta=$const with an appropriate $\beta_n(\zeta)$ function. The thick line in Figure 5 consists of respective $\beta_n(\zeta)$ functions. providing the smallest possible value of ζ for any β. Part of the line is presented separately in Figure 6 and allows for specification of respective values of ζ^{min} and n for a given β. Thus, the present energy model provides a unique solution in the number of fragments and the length of friction zone within the intact structure leading to cracks occurrence. Bialas and Mróz[11] presented a more detailed analysis of the transition condition (31) considering also the energy of the film-interface subsystem.

Let us measure the value of specific surface energy Σ_c for a given crack line spacing l observed during a saturation stage, when further loading of the substrate does not induce additional cracking within the coating. The configuration is schematically presented in Figure 4. where a film of initial length L has segmented into n fragments. It holds $L=nl$. By knowing the number n of strip fragments we can put an upper and a lower limit on the non-dimensional parameter β, being a function of mechanical and geometrical parameters of the system. By substituting $L=nl$ into the definition of β, equation 34(c), we have

$$\beta_n^{\text{min}} < \frac{24h_f^2 E_f}{(nl)^3 \tau_c^2} \Sigma_c < \beta_n^{\text{max}} \qquad (37)$$

where β_n^{min} and β_n^{max} are given for any number of strip fragments and can be read from Figure 6. Thus, inequalities (37) put an upper and a lower limit on the value of one material parameter (e.g. Σ_c), when the crack spacing l at the saturation stage as well as other mechanical and geometrical data (e.g. τ_c, E_f, h_f) are known.

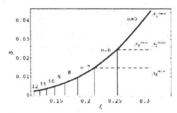

Figure 6. Value of β versus the length of friction zone ζ and the number of strip fragments n for $\lambda=0$.

Table 1. Experimental crack density ρ at saturation in SiO$_x$ films with different thicknesses h_f deposited on PET substrates (after Leterrier et al.[5]).

h_f [nm]	ρ [μm^{-1}]
30	0.725±0.087
53	0.5±0.022
96	0.336±0.015
103	0.389±0.021
145	0.377±0.027
156	0.362±0.013

SPECIFICATION OF FRACTURE ENERGY OF SILICON OXIDE THIN COATING

The value of $\tau_c=70$ MPa will be used for the specification of fracture energy of SiO$_x$ coating on PET substrate[4-6,9]. The other material parameters to be used in the calculations are the Young moduli of the film and the substrate. Further calculations will be carried on with the values $E_f=73$ GPa, $E_s=4.7$ GPa[3,4,6-9]. Experimental mean crack spacing l or crack densities at saturation in SiO$_x$ films with different thicknesses deposited on PET substrates are given in Table 1. The data were obtained by Leterrier et al.[5] and will serve as an input for measuring the value of fracture energy Γ by means of inequalities (37). In all cases the thickness of the PET substrate was 12 μm. The relation between the mean crack spacing l and the crack density ρ has the form $l=1/\rho$.

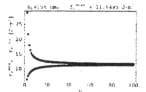

Figure 7. Upper and lower limits put on Σ_c for various film thicknesses h_f as a function of n (using the experimental data in Table 1); $\Sigma_c^{mean}=0.5(\Sigma_c^{max}+\Sigma_c^{min})$ – mean value of Σc obtained for $n=100$.

Figure 7 presents the upper Σ_c^{max} and the lower Σ_c^{min} limits put on Σ_c and obtained using inequalities (37) and the experimental data in Tables 1. It is seen that the values of Σ_c^{max} create a decreasing sequel, whereas the values of Σ_c^{min} create an increasing sequel. Thus, the bound specified for n fragments falls within the range defined by Σ_c^{max} and Σ_c^{min} for n-1 fragments. The mean values of Σ_c defined by $\Sigma_c^{mean}=0.5(\Sigma_c^{max}+\Sigma_c^{min})$ can be related to the film fracture energy Γ by the following equality $\Sigma_c^{mean}=2\Gamma$, where 2 indicates that two new surfaces are created for each segmentation crack.

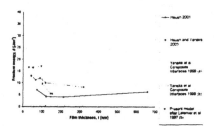

Figure 8. Fracture energy Γ as a function of film thickness h_f.

Figure 8 presents Γ as a function of film thickness obtained from $\Sigma_c^{mean}=2\Gamma$ for Σ_c^{mean} calculated for $n=100$. The results are obtained for experimental data sets presented in Table 1 and compared with data found in the literature. Yanaka[9] assumed a shear deformation either in the substrate or in the film in the shear lag model. Resulting film fracture energies Γ as functions of film thickness are indicated in Figure 8 by (a) and (b), respectively. Using fracture toughness $K_c=0.794$ MPa(m)$^{1/2}$ (Leterrier et al.[5]) for the glass film, the corresponding fracture energy becomes $\Gamma=K_c^2/2E_f=4.32$ J/m^2 which is in the range of Γ in Figure 8. It should be stated that both the present model and the values of Γ found in the literature and presented in Figure 8 are reasonably close in the range $4\div18$ J/m^2. Only Yanaka et al.[9] predict Γ to be in the range either $11\div35$ J/m^2 or equal approximately 0.5 J/m^2, depending on the assumption of shear deformation in the shear lag model (substrate or film, respectively).

CONCLUSIONS

The main assumption of the energy model is equal spacing between segmentation cracks in the saturation stage. Experimental results show a variation of crack spacing around a mean value. Using a stress redistribution model Agrawal and Raj[12] showed that the maximum and minimum crack spacing should differ by a factor of two. This effect cannot be captured by the present approach. There is another phenomena that is also not included in the present analysis. Since we use a topological transformation from an uncracked to a fully damaged state of the system, we do not consider the loading history - it is not possible to model a relation between applied load and crack density. This effect can be easily captured by stress redistribution models (see, for instance, Białas and Mróz[13]).

ACKNOWLEDGMENT

The financial support of EC (Contract No MERG-CT-2006-036548) is gratefully acknowledged.

REFERENCES

[1]T. Futatsugi, S. Ogawa, M. Takemoto, M. Yanaka, Y. Tsukahara, "Integrity evolution of SiO$_x$ film on poltyethylene terapthalate by AE characterization and laser microscopy", *NDT&E Int.*, **29**, 307-16 (1996).

[2]C.H. Hsueh, "Analyses of multiple cracking in film/substrate systems" *J. Am. Ceram. Soc.*, **84**, 2955-61 (2001).

[3]C.H. Hsueh, M. Yanaka, "Multiple film cracking in film/substrate systems with residual stresses and unidirectional loading" *J. Mat. Scie.*, **38**, 1809-17 (2003).

[4]Y. Leterrier, L. Boogh, J. Andersons, J.-A.E. Månson, "Adhesion of silicon oxide layers on poly(ethyleneterephtalate). I: Effect of substrate properties on coating's fragmentation process" *J. Polym. Scie. B: Polym. Phys.*, **35**, 1449-61 (1997).

[5]Y. Leterrier, J. Andersons, Y. Pitton, J.-A.E. Månson, "Adhesion of silicon oxide layers on poly(ethyleneterephtalate). II: Effect of coating thickness on adhesive and cohesive strengths" *J. Polym. Scie. B: Polym. Phys.*, **35**, 1463-72 (1997).

[6]A.P. McGuigan, G.A.D. Briggs, V.M. Burlakov, M. Yanaka, Y. Tsukahara, "An elastic-plastic shear lag model for fracture of layered coatings" *Thin Solid Films*, **424**, 219-23 (2003).

[7]M. Yanaka, Y. Tsukahara, Y. Nakaso, N. Takeda, "Cracking phenomena of brittle films in nanostructure composites analysed by a modified shear lag model with residual strain" *J. Mat. Scie.*, **33**, 2111-9 (1998).

[8]M. Yanaka, Y. Kato, Y. Tsukahara, N. Takeda, "Effects of temperature on multiple cracking progress of sub-micron thick glass films deposited on a polymer substrate" *Thin Solid Films*, **355-6**, 337-42 (1999).

[9]M. Yanaka, T. Miyamoto, Y. Tsukahara, N. Takeda, "In situ observation and analysis of multiple cracking phenomena in thin layers deposited on polymer films" *Composite Interface*, **6**, 409-24 (1999).

[10]J. Stockfleth, L. Salamon, G. Hinrichsen, "On the deformation mechanisms of oriented PET and PP films under load", **271**, 423-435 (1993).

[11]M. Białas, Z. Mróz, "An energy model of segmentation cracking of thin films", submitted to *Mechanics of Materials*.

[12]D.C. Agrawal, R. Raj, "Measurement of the ultimate shear strength of a metal-ceramic interface" *Acta Metall.*, **37**, 1265-70 (1989).

[13]M. Białas, Z. Mróz, "Crack patterns in thin layers under temperature loading. Part I: Monotonic loading" *Engng. Fract. Mech.*, **73**, 917-38 (2006).

EFFECT OF WITHDRAWAL SPEED ON THICKNESS AND MICROSTRUCTURE OF 8MOL% YTTRIA STABILIZED ZIRCONIA COATINGS ON INORGANIC SUBSTRATES

Srinivasa Rao Boddapati and Rajendra K. Bordia
Department of Materials Science and Engineering
University of Washington
Box 352120
Seattle, WA, 98195

ABSTRACT

In many applications, a ceramic coating is needed on metallic substrates. In the processing of solution based coatings, there are significant constraint induced stresses during drying, sintering and cooling from sintering temperature. 8mol% yttria stabilized zirconia (YSZ) coatings have been processed by dip coating glass and 316 stainless steel substrates with water based slurry containing 7.5vol% of nanometer sized particles. Coated steel samples were sintered in argon at 1000°C for 2h. The thickness of the green coating is a function of withdrawal speed (V), viscosity, density and surface tension of the slurry. The effect of withdrawal speed on green stage coating thickness and on the microstructure of sintered films has been investigated. The green stage coating thickness was found to be proportional to $V^{2/3}$ and is in good agreement with the Landau-Levich theory. Dip coated films are prone to cracking during both drying and sintering stages. Although intact green coatings can be obtained even at low withdrawal speeds (1.5-9 cm/min), no intact sintered coating was found after sintering. However, intact and well sintered coatings have been obtained in the withdrawal speed range of 18-36 cm/min.

INTRODUCTION

Yttria stabilized zirconia coatings are used as thermal barrier and environmental barrier coatings, and in sensors and solid oxide fuel cells due to low thermal conductivity, high oxygen ion conductivity and good corrosion resistance. Yttria stabilized zirconia (YSZ) coatings are processed by such as dip coating[1-2], slurry deposition[3-4], spraying[5], laser deposition[6], chemical vapor deposition[7-10], and sol-gel[11]. Among these techniques, dip coating is a simple and cost effective method to process ceramic coatings on substrates of different geometries. The challenge is to process crack free YSZ coatings on different substrates whose thermal expansion coefficient is significantly different from that of YSZ. In order to reduce thermal expansion mismatch induced residual stresses, coatings are made by slurries containing nanoparticles which reduce the temperature required for sintering the film to near theoretical density. The process parameters that govern the green stage film thickness are viscosity, density, surface tension of the slurry, acceleration due to gravity and the withdrawal speed. However, at high withdrawal speeds, the film thickness is independent of the surface tension and is determined exclusively by viscosity, density, acceleration due to gravity, and withdrawal velocity[12]. The main objective of this work was to understand the effect of withdrawal speed on the thickness of green stage YSZ films and microstructure of the sintered films. Green stage film thicknesses were measured by dipping the glass slides in the YSZ slurry, as described by Guillon et.al[13]. After drying, the coated slide was broken into two pieces and measured the thickness of the films using scanning electron microscope. The effect of withdrawal speed on microstructure was studied by dip coating the stainless steel substrates and sintering them at 1000°C in argon.

EXPERIMENTAL PROCEDURE

Both microscope glass slides (length: ~30mm; width: ~10 mm; thickness:~ 1mm) and 316 stainless steel (length: 30mm; width: 10 mm; thickness: 1.5 mm) were used as substrates. Both glass slides and steel substrates were ultrasonically cleaned in acetone and dried prior to coating. 8mol% yttria stabilized zirconia (8YSZ) particles (TZ-8Y,Tosoh) with a primary particle size less than 100nm (Fig.1) were used for the preparation of slurries. The particle size of as-received powder was measured by dispersing 1wt% of the powder in isopropanol and ultrasonicating it for 40s. The volume fraction of YSZ slurry used for dip coating was kept at 7.5%. The flow chart depicting different steps involved in preparing YSZ slurry is shown in Fig.2. Darvan 821A (5wt% of the solids) and Polyethylene glycol (20M compound; 10wt% of the solids) were used as dispersant and binder, respectively. After mixing, the slurry was ball milled for 18h. After ball milling the pH of the slurry was adjusted to about 10.8 with NH_4OH. Withdrawal speed of the dip coater was varied in the range of 1.5 cm/min to 36 cm/min to vary the film thickness A dual speed dip coater and an Instron (Model: 5500R) machine used for testing the mechanical properties of materials were used for dip coating. The advantage of using a mechanical tester such as Instron is that the withdrawal speed can be varied over a wide range (0.1-500 mm/min) unlike the normal dip coaters. .

Viscosity of the slurry was measured as a function of shear rate using a viscometer. The viscosity of the slurry exhibited shear thinning behavior, as shown in Fig. 3. After dip coating the films were dried at the room ambient. The films on glass slides were broken into two pieces and measured the thickness of the films in a SEM (JEOL JSM 7000). The films on stainless steel were heated at 3°C/min in air up to 600°C, and then onwards in argon up to 1000°C, held there for 3h and cooled to room temperature at 3°C/min. After sintering, the films were examined in SEM.

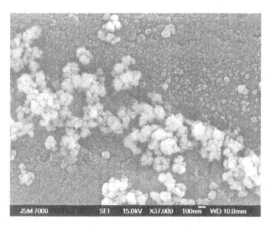

Fig.1: As-received 8mol% YSZ particles. The primary particle size is less than 100nm.

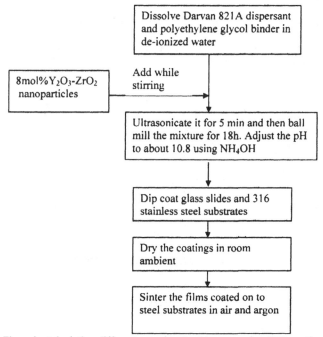

Fig.2: Flow chart depicting different steps involved in processing YSZ coatings

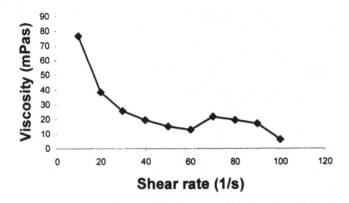

Fig.3: Viscosity of 7.5vol% YSZ slurry as a function of shear rate. The slurry exhibited shear thinning behavior.

RESULTS AND DISCUSSION

Effect of Withdrawal Speed on Green Film Thickness:

The thickness of the green YSZ films on glass slides as a function of withdrawal speed is shown in Fig.4. The thickness of the films increased with the withdrawal speed.

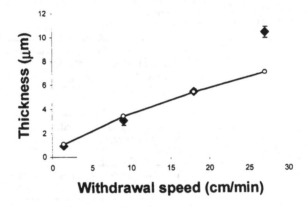

Fig.4: Effect of withdrawal speed on the thickness of 8mol%YSZ coatings on glass slides (solid diamonds are experimental data; open circles are data fit using equation (1))

Landau and Levich[12, 14] developed theoretical understanding to estimate the thickness of a liquid film remaining on the surface of a solid body withdrawn from quiescent liquid. Two regimes have been identified:

$$\text{When } V \ll \frac{\sigma}{\mu}, \qquad h \approx 0.93 \left[\frac{\mu V}{\sigma} \right]^{1/6} \left[\frac{\mu V}{\rho g} \right]^{1/2} \qquad (1)$$

$$\text{When } V \gg \frac{\sigma}{\mu}, \qquad h \approx \left[\frac{\mu V}{\rho g} \right]^{1/2} \qquad (2)$$

Where, V: withdrawal velocity, μ, ρ, and σ are viscosity, density and surface tension of the liquid, respectively, and g is acceleration due to gravity. The assumptions in the model are: (1) the liquid is Newtonian, (2) the effect of gravity is negligible compared to surface tension, (3) the substrate is perfectly planar, and (4) no evaporation takes place. At lower withdrawal speeds, the film thickness is proportional to $V^{2/3}$ whereas at higher withdrawal speeds the film thickness varies as $V^{1/2}$ as surface tension has no effect and the capillary number ($\mu V/\sigma$) approaches 1.0. In the present work, for the purpose of calculating σ/μ, the surface tension of the slurry was assumed to be around 0.1 N/m and the viscosity of the slurry was measured to be 25 mPas (average of viscosity values at different shear rates). The withdrawal speeds used (1.5-27 cm/min) are much smaller than the σ/μ ratio (4 m/s). Therefore, regime (1) of the Landau-Levich equation applies to the present work. Since the properties of the slurry, such as surface tension and density, were not measured, equation (1) was evaluated at one of the experimental data points to determine the constant in equation (1). The solid line through open circles in Fig.4 is fitted using equation (1) and compared with the experimental data. The data fits quite well, especially at lower withdrawal speeds, and confirms that the variation of film thickness as a function of withdrawal speed is in accordance with regime one of Landau-Levich equation.

The cross-sectional view of the green film coated at 18 cm/min is shown in Fig.5. The film is uniform and well adhering to the substrate.

Fig.5: Cross-sectional view of YSZ coating on glass slide (withdrawal speed: 18 cm/min)

Microstructure of YSZ Coatings on 316 Stainless Steel Substrates:

The macro- and microscopic observations on the green stage and sintered films are summarized in Table I. Although intact green films were obtained at 1.5 cm/min and 9 cm/min withdrawal speeds, after sintering no intact films were found in these two cases. Therefore, there exists a critical green stage film thickness below which no intact films will be formed when they are sintered. In the present case, the withdrawal speed corresponding to the critical thickness lies between 9 cm/min and 18 cm/min. The microstructure of the coating made at 27 cm/min withdrawal speed is shown in Fig. 6. Although the film contains some porosity, the film sintered well even at 1000°C as the primary YSZ particles are less than 100 nm. Fig.7 shows a sintered particle agglomerate which did not break down during ball milling. These agglomerates impair the coating integrity and mechanical properties. The incidence of these agglomerates can be reduced by increasing the milling time.

Table I: Summary of macroscopic and microscopic observations of the coatings before and after sintering at 1000°C for 3h in argon

Withdrawal speed, cm/min	Green stage	After sintering
1.5	Film was semi-transparent and intact	No intact film; Patches of particles were seem
9	Film intact	No intact film; Patches of film were seen on the substrate
18	Film intact	Crack free film
27	Film intact	Crack free film
36	Film intact	Crack free film except at the bottom edge where the coating cracked and debonded

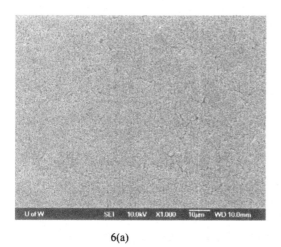

6(a)

6(b)

Fig.6: Microstructure of 8mol%YSZ coatings on 316 stainless steel made at 27 cm/min withdrawal speed: (a) top view of the coating, and (b) high magnification view of microstructure of the coating sintered at 1000°C showing well sintered structure.

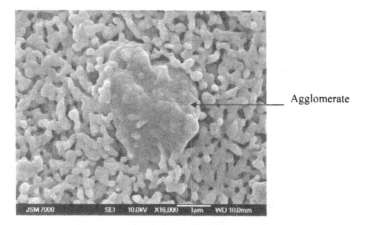

Fig.7: An agglomerate in sintered 8mol% YSZ coating (withdrawal speed: 36 cm/min)

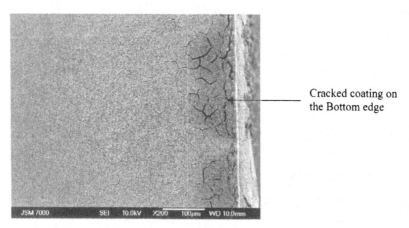

Fig.8: Cracking of 8mol% YSZ coating on the bottom edge. The thickness of the coating on this edge is higher than the thickness of the coating on other areas of the substrate as the slurry settled down to this location due to gravity.

It was observed that at high withdrawal speeds a patch of thick coating layer developed at the bottom edge of the substrate where the slurry settles down due to gravity once the substrate is withdrawn. Fig.8 shows such a region at the bottom edge of the 36 cm/min withdrawal speed sample. The mud cracking observed in Fig.8 are typical of drying induced cracks. These cracks further open up during sintering and/or cooling from sintering temperature. It was reported earlier[2] that the coatings crack during drying when the thickness of the film exceeds a critical value. Although the film was intact on all other areas, the cracking of the film on the bottom edge is a clear indication that the thickness of the film exceeded the critical value. In fact, the film at this region chipped off after about 24h from the time it was cooled to the room temperature due to residual stresses induced during cooling by the thermal expansion mismatch. These stresses have to be kept to the minimum in order to increase the service life of these coatings.

SUMMARY AND CONCLUSIONS

The effect of withdrawal speed on green stage YSZ coatings and on the microstructure of the sintered YSZ films has been studied by dip-coating glass slides and 316 stainless steel substrates.

The thickness of the green films increased with withdrawal speed and it was proportional to $V^{2/3}$ and in accordance with Landau-Levich model for liquid films. There is a critical withdrawal speed below which a green intact film can be obtained but no intact sintered film can be obtained after sintering.

No major cracking was observed either during drying or after sintering. The microstructure of the coatings consists of an interconnected network of pores and well sintered networks of YSZ particles.

ACKNOWLEDGEMENTS

The financial support for this work was provided by the Department of Energy (Grant number: DE-FG36-05GO15018), USA.

REFERENCES

[1]Y. Zhang, J. Gao, D. Peng, M. Guangyao and X. Liu, "Dip-coating thin yttria-stabilized zirconia films for solid oxide fuel cell applications," *Ceramics International*, 30, 1049-1053 (2004).

[2]S. Boddapati, L. Q. Nguyen, and R. K. Bordia, "Processing of 8mol% yttria stabilized zirconia coatings on 316 stainless steel substrates by dip coating," Innovative Processing and Synthesis of Ceramics, Glasses and Composites, *Proceedings of Materials Science & Technology (MS&T) 2006: Processing*, pp.413-419.

[3]G. Dell'Agli, S. Esposito, G. Mascolo, M. C. Mascolo and C. Pagliuca, "Films by slurry coating of nanometric YSZ (8mol% Y_2O_3) powders synthesized by low-temperature hydrothermal treatment," *Journal of the European Ceramic Society*, 25, 2017-2021 (2005).

[4]M. Gaudon, N. H. Menzler, E. Djurado, and H. P. Buchkremer, "YSZ electrolyte of anode-supported SOFCs prepared from sub micron YSZ powders," *Journal of Materials Science*, 40, 3735-3743 (2005).

[5]D. Perednis, O. Wilhelm, S. E. Pratsinis, and L. J. Gauckler, "Morphology and deposition of thin yttria-stabilized zirconia films using spray pyrolysis," *Thin Solid Films*, 474, 84-89 (2005).

[6]B. Hobein, F. Tietz, D. Stöver, and E. W. Kreutz, "Pulsed Laser Deposition of Yttria stabilized Zirconia for Solid Oxide Fuel Cell Applications," *Journal of Power Sources*, 105, 239-242 (2002).

[7]E. Martinez, J. Esteve, G. Garcia, A. Figueras, and J. Llibre, "YSZ protective coatings elaborated by MOCVD on nickel-based alloys," *Surface and Coatings Technology*, 100-101, 164-168 (1998).

[8]S. Chevalier, M. Kilo, G. Borchardt and J. P. Larpin, "MOCVD deposition of YSZ on stainless steels," *Applied Surface Science*, 205, 188-195 (2003).

[9]S. P. Krumdieck, O. Sbaizero, A. Bullert and R. Raj, "YSZ layers by pulsed-MOCVD on solid oxide fuel cell electrodes," *Surface and Coatings Technology*, 167, 226-233 (2003).

[10]R. Tu, T. Kimura and T. Goto, "High-speed deposition of yttria stabilized zirconia by MOCVD," *Surface & Coatings technology*, 187, 238-244 (2004).

[11]S.-G. Kim, S. W. Nam, S.-P. Yoon, S.-H. Hyun, J. Han, T.-H. Lim, and S.-A. Hong, "Sol-gel processing of yttria-stabilized zirconia films derived from the zirconium n-butoxide-acetic acid-nitric acid-water-isopropanol system," *Journal of Materials Science*, 39, 2683-2688 (2004).

[12] V. G. Levich, Physicochemical Hydrodynamics, Prentice-Hall, Inc., pp.675-683 (1962).

[13]O. Guillon, L. Weiler and J. Rödel, "Anisotropic microstructural development during the constrained sintering of dip-coated alumina thin films," Submitted to the Journal of the American Ceramic Society.

[14]L. D. Landau and V. G. Levich, "Dragging of a liquid by a moving plate," *Acta Physicoch. USSR*, 17, 42-54 (1942).

Author Index

Author Index

Printed in the United States
By Bookmasters